Managing Mathematically

Terry Green

and

John Webster

M

Macmillan Business, Management and Administration Series

First published 1976 by
THE MACMILLAN PRESS LTD
London and Basingstoke
Associated companies in New York Dublin
Melbourne Johannesburg and Madras

SBN 333 19113 7

Text set in 11/12pt Photon Baskerville, printed by photolithography and bound in Great Britain at The Pitman Press, Bath

Editor's Introduction

Alan Hale

This series is a direct outcome of the opportunities and challenges created by the rapid expansion of higher and further education in the past decade.

The expansion involved changes in the structure of advanced education, through the CNAA, the Polytechnics, the Regional Management Centres and the professional bodies which encouraged staff to develop new and experimental teaching. Substantial changes have taken place in the definition, scope and methodologies of the social, administrative and management sciences leading to modifications in the presentation of these subjects. Many new full-time students and staff have questioned traditional approaches and methods and have established more open discussion and debate on their courses. Demands for qualified manpower led to an expansion in part-time education and increased questioning by students in full-time jobs of the relevance of their studies.

Each of these developments has had a profound impact on the structure and content of courses and given fresh impetus to the discussion and modification of curricula and teaching methods in polytechnics, universities and colleges of further education. The editor and authors of the books in this series have made a deliberate attempt to respond to these changes.

The books set out to provide a comprehensive and up-to-date introduction to the ideas and methods of their subjects for specialist and non-specialist students in fields such as business and management studies, social science and administration. Their aim is to help students who have little or no previous knowledge of them to achieve a mastery of the scope and basic techniques of their subjects and to use them critically and with imagination for further study or for pratical professional applications. They also seek to make some contribution to discussions of teaching and learning problems in their field.

Many introductory books present their subjects as a coherent body of knowledge of which the logic is self-evident and the concepts and methods clear to the careful reader. Students do not always find this so. Confronted as they are by a well-established discipline which has developed a particular method that may not bear any obvious relationship to the way in which they have been accustomed to think or to realities as they see them, students often have difficulty in comprehending the significance and detail of the forms of analysis it employs.

The editor and authors of the series felt that they should not take for granted the 'self-evident logic' of their subjects, but try to demonstrate to readers the ways in which their disciplines provide an effective framework for the analysis of problems in their field. When abstractions or concepts are introduced their functions and limitations are explained. Where methods or techniques are described the authors show why they take the form they do and the ways in which they may be used for particular tasks of analysis.

Students often criticise courses because their subjects or parts of them do not have any obvious, or immediate, practical applications. They may present what appear to be unnecessarily complicated ways of dealing with quite straightforward problems or, paradoxically, they may be regarded as over-simplifying or ignoring difficulties which are experienced in real situations. Criticisms of this sort are due to misunderstandings about the nature of subjects. Systems of knowledge provide generalisations which are derived from a variety of abstractions and models. Some of these yield tools of description and analysis that have direct applications. Others suggest ways of looking at problems that, however fruitful, may have only limited or indirect applications.

The authors have tried to make clear the relevance of their subjects. Where concepts and methods have direct applications they show how, and under what circumstances, they can usefully be applied. Where they are of indirect use they show how a process of simplification may isolate and draw attention to the important characteristics of a complex problem, or how the study of complicated or abstract aspects of a problem may throw fresh light on it.

The authors have rejected a view, reflected in many basic textbooks, that students at an introductory level should concentrate on 'learning up' the information and techniques of their subjects and not be troubled with discussions of concepts and analytical method. This 'descriptive' approach, which divorces the study of techniques from that of conceptual and analytical structures, makes it more rather than less difficult for students to appreciate how a technique has developed, why it takes a particular form and how it functions. As a result students can spend a considerable time trying to understand, with limited success, a method of describing and presenting information or a method of analysis and fail to achieve any real facility in using it. The discussion of concepts and analytical method also acquaints a student with some of the difficulties and controversies surrounding the ideas and techniques he is studying. Without such knowledge he is unlikely to appreciate their limitations or establish any real ability to dscriminate between alternative approaches and methods.

One of the more important aims of education is to develop a student's capacity to formulate and solve theoretical and practical problems. It is clear that few business and administrative problems are in practice separable into the neat categories represented by disciplines such as economics, accounting, law, sociology, psychology and computing. But most courses are based on combinations of studies in these and similar discrete disciplines which are rarely effectively integrated. It is recognised that the development of bodies of knowledge which provide rigorous rather than superficial integrative approaches will be a long and difficult task. The editor and authors of the books in this series are aware of this problem and within their limitations have attempted to indicate points at which contributions from other disciplines are necessary to the analysis of the problems with which they are dealing. It is thus hoped that in the long run the series will make some contribution to the development of interdisciplinary approaches.

The problems outlined above are common in the teaching and learning of many subjects which, although emerging historically as systems for analysing and solving practical problems, have developed advanced methodologies and a logical order of presentation that may not bear an obvious relation to the practical problems with which they are supposed to be concerned.

Although mathematics has long been used in the analysis and solution of scientific and technical problems and is now applied in a wide variety of management activities many still regard it as incomprehensible or irrelevant. Previous experience of mathematics has led some to believe that it is difficult and dull subject in which complicated techniques are used to solve problems that are either too trivial or too advanced to interest them. Others see little need for it in their work where well-tried non-quantitative methods appear to be quite adequate.

It is surprising how often those who avoid mathematics do in their daily experience estimate variables, assess risks and play games which involve the sort of mental activities with which mathematics is concerned. Perhaps difficulties arise because mathematics uses unfamiliar symbolic language to deal with familiar situations and that it employs rational and logical approaches to problems that are often dealt with in a fairly intuitive way.

The power of mathematics is that it enables significant characteristics of problems to be identified more precisely and represented more simply than can be done by non-quantitative means. Problems in management and administration are increasingly dealt with by the use of symbolic and quantitative methods and learning difficulties in subjects such as economics or accounting often arise because of a lack of basic mathematical education. The authors of this book have tried to present mathematics in an interesting and relevant way and it is hoped that it will help many who are uncertain about the subject to develop an ability to use basic mathematics with ease and confidence.

The editor and authors of the books in this series are conscious of their limitations in attempting to implement their ideas in writing and teaching and do not suppose that their presentation will solve students' learning problems. They do not ignore the critical importance of motivation and sustained and disciplined study as factors in effective learning. But they felt that if subjects were presented in a way that made their form and justification explicit rather than implicit this would aid teaching and learning.

In seeking to achieve their aims the books in the series have been subjected to a great deal of critical scrutiny. Each is written by more than one author. This has enabled authors to combine a comparison of views with a considerable, and sometimes uncomfortable, degree of mutual criticism. The editor and authors have all, in recent years, had considerable experience of designing and teaching new CNAA honours degree, diploma or professional courses. Their manuscripts have been discussed with colleagues in education and the professions and have been tested in classes with students.

My thanks as editor are owed to the authors who responded to my request to write the books in the form that I have outlined. This has involved them in an extremely demanding process. The fact that we shared some basic assumptions about education and learning was of great help. The editors and staff of Macmillan with whom we have worked have shown great patience and could not have done more in difficult circumstances to encourage the series. My thanks are also due to George Brosan, Maurice Peston and Bruce May, who through many discussions have significantly influenced my educational ideas, and to my wife Diane, who has kept a discerning eye on my activities and the progress of the series.

February 1976 ALAN HALE

Contents

Preface to the Student

Why You Need This Book

You need this book to help you to BE MATHEMATICAL. Surely it is no longer respectable for a businessman to say in a jovial and rather complacent fashion 'of course I know nothing about all this mathematics business . . .'. The increasing use of mathematics techniques and arguments in business means that at the very least every businessman should know enough to appreciate the contribution mathematics can make, and the effect it has, on his management role. But more than that, he needs to be numerate enough to participate in this mathematical activity in order to derive the full benefits from it.

This book has been written to help you develop mathematically and so reach a higher level of numeracy. We will not ask you to perform mathematical tricks even more elaborate than those you have learnt already at school. Our object is not to make you into a MathematicIAN; we want you to be mathematiCAL. Being mathematical is an attitude of mind, nothing to do with being especially adept at adding and subtracting, and enables you to be analytical in your judgement of problem situations.

For example: a queue in a supermarket would be seen in terms of colours, shape and their interaction. From the mathematical point of view the same queue would be seen quantitatively in terms of the number of people in the queue, the rate at which they are served and how the two are related.

How to Read This Book

Unlike many mathematics books (see Chapter 9) this one is to READ, starting from page 1. To read this book you need:

a pen
a lecturer
a library

Mathematics is an active pursuit, like swimming. You would not expect to be able to swim after reading a book; you need to try out the instructions and through these to learn new skills. So here and there in the text we ask you to pick up your PEN and DO mathematics.

Your mathematical progress is marked by the questions of increasing complexity you ask yourself. We shall ask you questions directly and suggest others for you to ask your lecturer. The answers will not always be found within the text, for the book was not intended to replace a lecture course, but to be used in conjunction with it. Inevitably you will have difficulties and you need a LECTURER to help you with them.

All the material in the text is designed to help you develop mathematically. It is a selection of the mathematics and statistics which is used extensively in business and should equip you to study further from other books. You need a LIBRARY to investigate in greater depth topics that have been introduced in this book, and to find out about ideas and techniques we have not included.

Your Examinations and You

Your mathematical progress in college will be measured by the success you have in examinations. This is a problem that you cannot ignore. Chapter 10, Examination of Examinations, demonstrates how examination questions should be interpreted enabling you to recognise what the examiner expects to be included in the solutions.

Synopses of Chapters

Preface To the Student

Preface To the Lecturer

1 WHAT'S YOUR PROBLEM?
Using real situations, the progression is made from solving the well-determined undiscipline problems at school to solving the complex multidiscipline problems of the business world. From this is demonstrated the needs of the businessman if he is to operate efficiently and effectively, and hence the *raison d'être* of the book.

2 DATA: WHAT CAN IT SAY?
A problem situation is used to demonstrate the need for data collection. The methods of interpreting data are grouped into five levels, including visual, statistical measures and probability. The contribution each successive level can make to the understanding of behaviour in the problem situation is assessed.

3 LINE UP THE FUNCTION
The straight-line model arises naturally from the business environment and is investigated both from the mathematical and statistical point of view. The idea and notation of 'function' is developed from the study of the line and its business applications. These ideas are used to introduce the idea of slope and its invariance under change of scale. A further extension gives a business meaning to the solution of equations.

4 FUNCTION EXTENDED
Slopes of curves are studied in detail and the notation of differential calculus is introduced.

Calculus ideas are developed to solve maxima and minima problems, particularly in the case of finding the best-fitting straight line to a set of data. Finally Newton's method of solving equations using an iterative method introduces the student to algorithms, flow diagrams and rates of convergence.

5 MODELLING SYSTEMS
The behaviour described by an equation and inequality is discussed. The same system is represented by models of varying complexity by making different assumptions. The difficulties of obtaining the data for each model are considered. The solution to a linear programming model is found and its validity investigated.

6 MODEL ANALYSIS
The choice of a model is discussed by introducing the concept of total error. The simulation approach to problem-solving is described, illustrated by a simplified but real problem, using one of the models of Chapter 5.

7 COPING WITH UNCERTAINTY
Uncertainty is discussed in terms of probability and the interpretation of collected data. A quality-control problem is used to develop the statistical ideas of risk and significant difference. Stress is placed on the generality of the ideas of statistic, parameter and distribution, rather than on the study in detail of any one particular probability model.

8 BE SURE OF THE DATA
Various agencies collect data and present the results in condensed form to decision-makers. This chapter discusses the problems of interpreting such results, planning successful data collection projects and using mathematical specialists and their techniques effectively.

9 AFTERMATHS
The reader is invited to examine his mathematical attainment and to consider ways of continuing development. Attention is drawn to topics not covered in the text and advice is given on how to deal with them. Some useful literature is listed.

10 EXAMINATION OF EXAMINATIONS
Most examination questions in quantitative methods are artificial exercises in mathematical and statistical techniques and as such should be relatively easy to identify and solve.
Examination questions which have been set on business diploma and degree courses are discussed and solved in detail. The solutions are designed to bring out the underlying mathematical principles, explore the relevance of the assumptions and give the effective methods of solution under examination conditions.

To the Lecturer

The Style of the Book

This book looks quite different from most other basic texts in mathematics and statistics, which have distinct chapters on surveys, significance testing, sampling, the Normal distribution, the straight line etc. . . . These areas are included in this book, but not separately. The approach used is to consider realistic problems and to introduce the appropriate mathematical technique as the need arises. We believe this helps the student to understand the advantages of looking at problems mathematically, and why in a particular situation one technique is used in preference to another.

The Book and the Student

The book is complementary to, not a replacement for, the lectures and students will benefit from reading it throughout the course. It is designed to develop a mathematical awareness in the student, encouraging him to become involved in his learning situation and to approach your lecture material in a questioning frame of mind.

1
What's Your Problem?

The purpose of this book is to show how mathematics and statistics can be used in solving problems. But before any attempt at solution can be made, the nature of the problem situation must be closely investigated.

Nature of a Problem

A problem is a set of circumstances requiring settlement or treatment. It is not the circumstances themselves that constitute a problem but rather their effect. The same set of circumstances may produce several problems for different individuals depending upon their situation and purpose. Rain, for example, for the walker causes his path to become muddy and his clothes wet; to the motorist it brings poor visibility and slippery roads; for a farmer in England it may ruin crops; for a farmer in a hot climate it is no problem, but provides a solution to drought.

Sometimes it is not clear that there is a problem. A problematic set of circumstances may exist but it is possible to be ignorant of the effects or even of the set of circumstances itself. A person inside a heated room may be unaware of the cold outside, but if he were to go out the cold would be a problem to him. A sunbather on a beach is very much aware of the heat but may not realise the effects it can have on him – that is, sunstroke and sunburn. In both these cases information is needed to bring the problem to notice.

Once a problem is known, attempts can be made to solve it. However, usually there is no unique solution: just as one set of circumstances can produce several problems, so each problem may have many solutions, or it may not be solvable at all. For instance, someone about to take a walk in the rain may carry an umbrella or wear a raincoat. He may even choose to ignore the problem and get wet. It is clear that he must know about the umbrella or raincoat before he can select it as a solution. Greater knowledge therefore implies greater choice of solution. Experience of similar problems and their solutions is also an asset in deciding how best to treat the one in hand: someone having experienced only light showers may find difficulty in protecting himself against torrential rain.

Each person then has a range of solutions dependent upon his knowledge and experience. The particular solution chosen depends on the objective: if the walker's aim is to reach a destination then he might choose to take a taxi, but this would obviously be nonsense if his aim is to take a stroll. In this case, he might stay indoors and postpone the walk until later, in the hope that the rain might stop. Sometimes no solution is available at all and the problem must be endured: if the walker is already outside and far from shelter then he has no alternative but to get wet!

There is, therefore, a great deal to be considered before any problem is approached. Is the problem itself fully realised? What difficulties does it pose? What solutions are available and which is the best one to choose? As an aid to considering these questions a problem picture can be drawn which categorises all problems.

Picture the Problem

A problem situation has two complementary parts: the problem and the solution. There are two types of problems: those you know you have and those you don't know you have but which must be brought to your notice by investigation.

The solution part is similarly split into the solutions you can work yourself and those which are beyond your skills. The diagram below represents all problem situations.

		Solution	
		Can do	Can't do
Problem	Known	1	2
	Unknown	3	4

Problem picture

Any real-life problem belongs to one of the categories 1, 2, 3 and 4. To be an efficient problem solver it is necessary to establish procedures to deal with each type of situation.

Situation Type 1

Everyone should be familiar with situations of type 1 since the school classroom is the common habitat. The problem is known because it is presented by the teacher; it can be solved because the teacher has provided the method. An example would be to find the area of a rectangle given the formula

$$\text{Area} = \text{Length} \times \text{Breadth}$$

and some values for the length and breadth. Examples of problems from school are perhaps not the best to choose since they are artificial in the sense that you yourself don't really have the problems. They are set in order to provide examples of working out formulae or using techniques; there are no consequences to reaching a solution. At this point the exercise book is shut. The value is in providing experience and techniques so that if you were to meet the problem of finding an area, or something similar, you would have a method of solution.

An example from real life of problem situation type 1 would be of two people buying blocks of notepaper and trying to establish the better buy and hence the place to purchase from in future. Bill pays 20p for a block of 150 sheets and Ben pays 15p for a block of 100 sheets. Employing the techniques of simple division you can find that Bill is paying 0·14p while Ben is paying 0·15p per sheet, so it is clear that Bill has the better buy. Of course, assumptions have been made in reaching this conclusion: we assume that the sheets are of equal size and quality and that the two places of purchase could be reached with equal convenience. If any of these were not true then the decision reached may not be valid. Different results could be argued from different points of view: cost, quality, size and ease of purchase. The criterion of judgement could vary from person to person according to the relative importance of these characteristics.

Situation Type 2

Unfortunately we meet type 2 situations frequently and it is not hard to find an example. If you own a car and it won't start, then that is a problem and you certainly know about it! You can't solve it because you don't have the expertise, so you must seek outside help and call in the services of a motor mechanic. Once help is obtained then you're back to situation type 1 once more: you know of a problem which you (with help) can solve. For success in all four situations it is necessary to convert them to situation type 1.

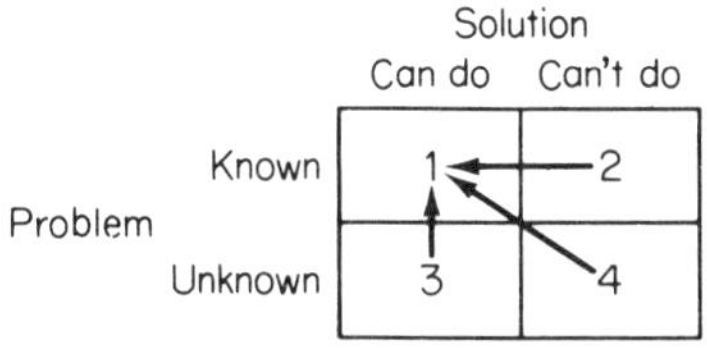

Converting situations

Situation Type 3

Suppose you have a current account at a bank but don't keep records of the cheques you write; then you may be overdrawn (a problem) without being aware of it (an unknown problem). If you were aware of your predicament then hopefully you could solve it by depositing more cash. In order to convert this situation to type 1 some procedure must be set up to make the problem known. Information channels need to be created between you and your financial affairs to carry information on the state of those affairs. In this case a simple record of cheques would be sufficient. The channels fulfil a dual role: they monitor the progress of your activities, providing information on how things are running, and they highlight a problem when it occurs. In each case there is then an opportunity to take appropriate action. Of course it is possible to adopt a head-in-the-sands approach and be blissfully ignorant of any difficulties, at least for a time. But problems don't disappear just because they are unknown; sooner or later they make themselves felt, usually through someone else's information channel: either the bank's when at some future date you receive your statement of account, in red; or the shopkeeper's when he refuses to accept your cheque for goods you wish to buy. It is better to create your own channels tuned to your purposes than to rely on those of others. You can then sort out your problems at the time and in the way most convenient for you.

Situation Type 4

Situation type 4 is a simple extension of type 2 or 3. Both information channels to reveal the problem, and assistance to solve it, are required for conversion to type 1. To illustrate, consider a manager of a hotel who will be a recurrent figure throughout this text.

The manager controls a hotel of 1000 rooms and is responsible for all the services: clean sheets, food, lighting, heating, room service etc., that are offered. Each service relies on a whole complex of other activities. For instance, the clean sheets are taken from linen cupboards and arranged on the beds by chamber maids; dirty sheets are sent to the laundry by van where they are washed and dried; the clean sheets are delivered, again by van, to the hotel and stored in the linen cupboards. The cycling of sheets from the hotel to the laundry is illustrated in the diagram below. Each box represents a process through which the sheets pass.

One problem is to determine how many sheets are required in the system: if too many, then the money spent on them is more than is necessary; if too few, then customers will be turned away or guests will have to suffer a lower standard of service with sheets being changed less frequently.

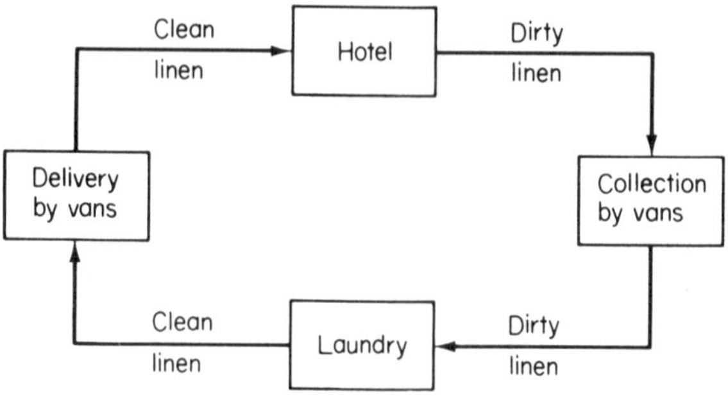

The linen cycle

The manager may not realise he has too few sheets (an unknown problem) until one day the linen cupboards are empty. On the other hand he may NEVER discover that the number of sheets is too great, instead flattering himself on giving good customer 'service'.

The solution to the problem is complicated; certainly no simple mathematical formula exists for solving it. If simplifying assumptions could be made, for example if the sheets are changed every day and it takes exactly 4 hours for them to be washed, then they could be back on the beds the same day and the simple formula

$$\text{number of sheets required} = 2 \times \text{number of guests}$$

would suffice. Unfortunately, in practice the laundry would take longer than that and extra sheets would be needed to make allowances for those being laundered, collected or delivered. The solution requires a complete analysis of the linen cycle. A discussion of an approach to this is contained in Chapters 5 and 6.

Variability

The solution to the problem above depends on several factors: the number of guests staying, the length of their visits, the speed of the laundry service, and collection and delivery services. Weather conditions and the traffic situation will also have a bearing on the problem. Information channels could be set up to determine the behaviour of these factors. For example, as each guest arrived and also when he left a record could be made of the date. It would then be possible to determine how many guests stayed in the hotel each night of the year. This information would then be useful in calculating the number of sheets required. Matters would be relatively straightforward if the same number of people arrived every day and always stayed for the same length of time, e.g. 1 week, but unfortunately these figures vary. It is quite likely that figures for Saturday and Sunday will be high, that more guests stay in summer than in winter, that some people stay for 2 days, some for 1 week, 2 weeks, etc. . . . This variability makes the best solution to the sheet problem more difficult to find: 1000 sheets may be enough in winter, but not in summer; 5000 sheets may be sufficient for a weekday, but not for a weekend. One criterion of choice of solution is that the number of sheets held should satisfy 'most' circumstances. It is difficult to say what 'most' circumstances means; there is always the freak situation of, say, a football match in town which brings a sudden deluge of guests.

Each of the other factors also varies, and several can often work 'out of tune': the laundry might process very few sheets at the weekend when the number of guests is very high. Statistics, the science of variability, copes with these difficulties and Chapters 2, 7 and 8 discuss statistical ideas.

Effectiveness of Solution

We have so far presumed that once a problem was known and a solution obtained then the problem could certainly be overcome by implementing that solution. For example, if a shortage of sheets occurred at the hotel then the purchase of more would effectively combat the shortage. However, it can be seen quite clearly from the linen cycle diagram that holdups at any point in that cycle could cause a shortage. Suppose, for example, that the laundry is working to capacity. Then the purchase of more sheets cannot solve the problem, as the laundry cannot handle them. The solution lies in the increase of the laundry facilities: the machines or staff. The most obvious solution of purchasing more sheets is not always the correct one and the temptation to grab it needs to be overcome. For an effective solution, links need to be established between the symptoms of a problem (the shortage of sheets) and the cause of those symptoms: poor delivery service, inadequate laundry facilities etc. . . . or a combination of these. Each activity: laundry collection, delivery etc., is a candidate for causing failure of the clean-sheet service in much the same way as each link may cause the breakage of a chain. The linen service is a connected system of these activities and as such is best studied as a whole. A simplified model often helps.

A model describes certain aspects of behaviour of the system it represents and omits unimportant aspects like the colour of the sheets. The diagram of the linen cycle is a simple model: the direction of movement of the arrow round the paper is a copy of the behaviour of the sheets in their real-life journey round the cycle; each 'box' represents a process through which they pass. The model is a simplification of the real system, however: no allowance has been made for sheets which are stolen or discarded because of wear; there is no representation of the time taken to complete any of the processes; but, nonetheless, the model does give a picture of the system and indicates how the different parts are connected. In Chapter 5 the concept of a model is developed further.

Multidisciplinarianism

The dependence of a solution on many factors, the variability of these factors, the processes of obtaining information all serve to make a real-life problem rather more difficult to tackle than a school problem. There is yet one more difficulty: this is the multidisciplinary nature of most real-life problems.

At school, problems are put into compartments and given subject labels: mathematics problems, French problems, chemistry problems. Each subject is studied at a different time of day and perhaps in a different room. The solutions to a subject's problems lie almost exclusively within the boundaries of that subject. Only occasionally are the results of one discipline employed in another: for example, mathematics in physics or chemistry. This method of study leads the student to file away information on each subject into rooms in his mind. His knowledge is compartmentalised. He proceeds happily from room to room opening and closing doors and solving different problems. The result is a unidisciplinary approach to problem-solving, which is inadequate to deal with more complex real-life problems.

The complete solution to the linen problem involves a study of the transport system of sheet collection and delivery, the mechanics of the laundry, the behaviour of the guests and the cost aspects. This draws on many disciplines: statistics, psychology, transport and economics; each contributes information on one aspect of the problem and all the results must be co-ordinated to obtain the complete picture.

The Problem-solver

The aim of the businessman is to be an efficient and effective problem-solver and decision-maker. Most decisions contain a quantitative aspect and this book deals with this. A problem could be any one of the four types for the businessman. By gathering information on his day-to-day affairs unknown problems will become known. Chapters 2 and 8 explore the ideas of data collection.

The same problem may have many solutions depending on the businessman's objective. When deciding the number of sheets to have, is the objective to operate with as few sheets as possible, or is it to find a system that works as cheaply as possible, or a system as near to the present one as possible in order to cause the least disruption in the changeover? To be efficient, the businessman must first define his objectives. By expanding his own knowledge and skills he enlarges the range of solutions available to him. An objective of this book is to afford the prospective businessman a greater measure of numerical skills and a higher level of numeracy. In particular, Chapters 3 to 7 are directed towards this aim.

The single most important concept to be grasped is that of the model, which lends itself to the establishment of the cause of a problem and to promising changes that might be made. The model is the key to being an effective problem solver.

By enlisting the help of experts, solutions outside the scope of the businessman's skills can be attained.

In order to perform this function efficiently the businessman needs some appreciation of his own capabilities. He must realise how far he can proceed alone on a problem and when it is appropriate to call in the specialist. Furthermore, he must be acquainted with the approaches and the language used by the specialist, to their mutual advantage. At the outset of a study the specialist will need co-operation in familiarising himself with the problem. At various points in the study information will be required and the businessman may have to organise its collection. The specialist will submit reports throughout the project which the businessman must understand; these will contain any results reached so far and on occasions, when the study could take many courses, he must respond to these reports and direct the study a particular way, sometimes curtailing it completely.

Without this acquaintance with the specialist's work, the project takes its own course and he cannot reap the full benefit of the specialist's skills. The businessman's knowledge in this area is a measure of his control over the project.

A recurrent theme of the book is to consider what lies ahead of the material and techniques already discussed, suggesting generalisations of particular approaches. Chapter 9 in particular deals with the extension to new situations and new problems.

2

Data, What Can it Say?

Systems

Every system is constructed in an attempt to solve a problem. The provision of letter boxes at certain points in a town and the arrangements for the collection, sorting and delivery of letters makes up the postal system which is a solution to the communication problem. It is not the only solution: the telephone, telex and telegram are others. The system itself can be varied in many ways to cope with different circumstances: the number and location of letter boxes, the times of collection and delivery, the organisation of the sorting offices can each be varied and this is seen to happen on Sundays when demands on the service are light and at Christmas when they are heavy.

Systems, Behaviour and Data

Similar variations are possible with the linen system described in Chapter 1. The collection, washing and delivery of sheets is analogous to the collection, sorting and delivery of letters. In each case time is an important factor. The Post Office measures its level of customer service by how soon most letters reach their destination. The faster our hotel manager can arrange for dirty sheets to be laundered and returned to the hotel, the less sheets he needs. The analogy may be pursued further: both systems deal with an outside influence, the letter-writing public in one case and the hotel guests in the other. The system is designed to cater for the particular known behaviour of these outside influences. A good design relies upon a thorough knowledge of behaviour and this can only be achieved by an investigation of some kind. This would yield some data which would hopefully provide some information on behaviour; for instance, in order to find out how many nights guests stay in the hotel, you could simply choose a guest and ask him how long he's staying. This would be your investigation and his answer might be 'one night'. The data from your investigation would be

A GUEST STAYED FOR **1** NIGHT

Unfortunately, this piece of data doesn't say very much, because there is only one piece. The linen needs of this one guest during his stay may be worked out as one pair of sheets, but it is impossible to generalise and say how long other guests would stay and what their total linen requirements might be. Information on other guests is needed because the linen system is being designed for many guests, not just the one investigated.

To obtain this information, you could ask all the guests arriving on a particular day how long they intended to stay. This would yield much more data than the first investigation. However, the information contained in this data might not be accurate because some people might change their minds and stay longer or leave earlier than they first planned. This method of collecting data stands a good chance of giving a false picture of what really happened.

When collecting data the question should always be asked: 'Will this data give me the information I need?' It is advisable to think carefully of the alternative sets of data available and to choose the one which is the most likely to yield information that is accurate and reliable. In the case of the hotel, a possible source of data is the customer's hotel bill. On this is recorded, among other things, the arrival and departure date of the guest, and it is handed in for payment to the hotel's cashier when the guest departs. The bills are used for reckoning up at the end of the day and would then normally be thrown away. If arrangements could be made for them to be retained then all those relating to a particular arrival date could be sorted out and the length of visit of each guest who arrived on this date could be calculated.

The following figures give the length of visit of the 90 guests who arrived on Wednesday, 19 June:

Length of visit of 90 hotel guests who arrived on 19 June. Units – number of nights.

1	1	1	1	1	1	1	1	1	1
1	1	1	1	1	1	1	1	1	1
1	1	1	1	1	1	1	1	1	1
1	1	1	1	1	1	1	1	1	1
1	1	1	1	1	1	1	1	1	1
1	1	1	1	1	1	1	1	1	1
1	1	1	1	1	1	1	1	1	1
1	1	1	1	1	1	1	1	1	1
1	1	1	1	1	1	1	1	1	1

This table is very easy to interpret: every guest stayed for one night, the linen requirements are 90 pairs of sheets. If these guests are typical, then we might infer that all guests stay for one night and so predict their linen requirements. However, it it is reasonable to ask whether these guests are typical. Is the table above consistent with your expectations of the behaviour of hotel guests? If not, why not? From your background knowledge of hotel visits you might have expected some 2-, 3- and 7-day visits, but no such figures appear in the table. Possible explanations for the lack of these numbers may be:

(1) The reason for the visit of the 90 guests was to see a football match being held in that city on 19 June. Therefore they might not be typical of usual GUESTS at the hotel.
(2) There was a fire alarm or bomb scare on 20 June and all guests were advised to leave. In this case the DAY is not typical

Whatever is the real reason, it can be established and a different set of data obtained which is more representative of the hotel clientele:

Length of visit of 90 hotel guests who arrived on Saturday, 22 June. Units – number of nights.

1	3	4	6	9	1	2	1	1	2
2	7	1	1	14	7	3	4	8	3
8	2	7	2	3	2	1	3	7	8
1	2	8	2	5	4	7	14	2	7
3	14	2	1	8	1	5	2	9	2
4	2	1	7	2	4	8	7	2	4
1	7	1	7	10	12	1	2	14	8
2	1	8	14	1	2	3	2	7	7
9	3	7	3	1	8	2	5	2	1

This table is more difficult to interpret. Certainly there is information in there, somewhere, but it is not apparent. The problem is to handle the data in such a way that it conveys this information to us, and many techniques are available to do this ranging from the simple to the sophisticated. But rather than choose an elaborate method, in the misguided belief that this will give more information just because it is more elaborate, it is more sensible to start with the very simplest and progress to a higher-level technique only when the simpler is too crude to be of useful service.

Visual Interpretation

The simplest way to interpret data is to look at it and see if any characteristics of behaviour can be determined. Often this level of interpretation is sufficient as is the case with the data of 19 June. There is no sense in doing calculations when the behaviour pattern is staring you in the face. Just as, statistics aside, if you were asked whether an elephant were larger or smaller than a mouse (by someone who had never seen an elephant) you would not measure the animals. Unfortunately, people seem to lose their heads in statistical problems and do what in statistical terms amounts to just that.

Certainly from a visual investigation of the table above you can see that there are many visits of 1 and 2 nights and quite a few of 7 nights.

Apart from that it is hard to say very much so we need to pass on to the next level.

Organising Data

More information may be extracted by organising the figures. One way of doing this is to count how many guests stayed for each number of nights. This is called the frequency and tells how OFTEN (frequent) a particular length of visit occurred. This information is normally put into a table called a frequency table:

Frequency Table – Saturday, 22 June

Length of visit Number of nights	1	2	3	4	5	6	7	8	9	10	11	12	13	14
Number of guests staying for that length of visit (frequency)	18	21	9	6	3	1	13	9	3	1	0	1	0	5

The frequency table confirms the impressions obtained by the visual inspection of the data and furthermore it quantifies them. For instance, it is possible to say how many guests out of a total of 90 stayed for 2 nights: there were twenty-one, nearly a quarter of the total. It also provides a comparison of the popularity of various lengths of visit: certainly short visits are popular, a total of 39 guests, almost half of the total, stayed for 1 or 2 days. One-week visits are also quite popular and a few people stay for 2 weeks.

Once more the question must be asked: 'Does this data confirm my expectations? If not, why not?' This time the answer is 'Yes'. We know that people tend either to make short visits or they stay for a whole week. It might be surprising to see that such a high proportion of guests stays for 1 or 2 days rather than the full week. This is what we have learnt about behaviour from the investigation. Data is collected to find out what we don't know as well as to confirm what we do know. The reason for this particular behaviour could be that the hotel is in a large city and the guests are often businessmen rather than tourists. A hotel in a seaside resort might show quite a different pattern of behaviour with more of the guests staying for 1 or 2 complete weeks.

It is often helpful to put the frequencies on to a diagram to convey the pattern of the data because pictures have more direct impact than numbers. One such diagram is the bar chart. To construct this, bars are drawn and the length of each bar represents the value of each frequency.

This diagram reveals, at a glance, all that is contained in the frequency table. It is perhaps easier to compare heights on the

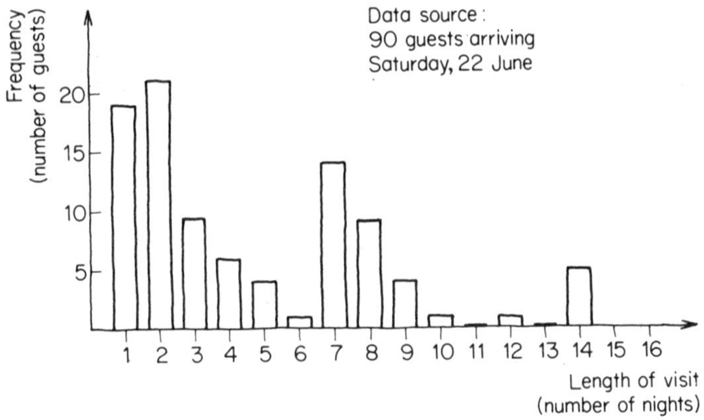

diagram than the numerical values of frequencies in the table. Several heights may be compared at once.

Different patterns of behaviour are shown up on the bar chart and in particular it is interesting to draw the bar chart corresponding to your expectations of behaviour before the real data were collected.

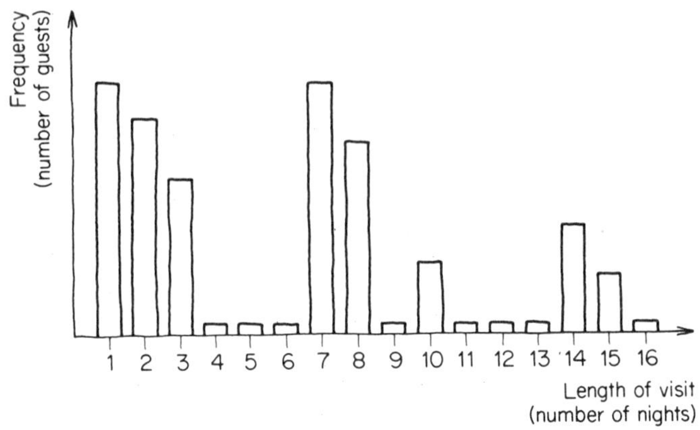

Expectations of behaviour

From the data collected on 22 June arrivals it is possible to work out the linen requirements of the 90 guests consistent with the level of service the hotel feels it would like to provide. For instance, if sheets are changed every day then a guest staying 3 nights needs 3 sets of sheets during his stay. If, however, the present system requires sheets to be changed every 4 nights, and also when a guest leaves, then this guest will require only 1 set of sheets; someone staying 5 days would need 2 sets. A table of sheet requirements may be drawn up for the different lengths of visit.

Table of requirements

Length of visit	1–4	5–8	9–12	13–16
No. of sets of sheets required	1	2	3	4

Using the table of requirements and the data of the table on page 12 the total number of sheet sets required is about 150. This is not to say that the hotel needs to have 150 sets clean AT ANY ONE TIME, because sheets can be laundered and used again. For instance, if a guest stays for 14 nights his bed is changed on the 5th, 9th and 13th day as a result of his visit, but the sheets removed on the 5th day will probably be ready to be put back on his bed on the 13th day. If sheets could be laundered and returned to the hotel in less than 4 days, only 2 sets of sheets would be needed: one set on the bed, the other in the laundry. Clearly, to proceed further with this problem the length of time taken to launder sheets must be investigated, and the following table contains the times (in hours) taken to launder 86 batches and return them to the hotel.

Time taken to launder each of 86 batches of sheets

105	89	78	64	129	82	74	69	68	110	76	81	95
70	102	24	28	81	125	19	94	65	70	82	93	135
19	96	72	94	116	89	86	88	115	99	58	119	80
53	119	69	89	127	84	47	127	70	74	64	69	64
111	45	80	67	93	63	79	82	87	87	110	84	130
64	123	135	83	113	94	86	125	68	138	95	115	120
71	86	72	63	79	126	71						

As before, the interpretation begins at level one, the inspection by eye. In answer to the question 'How long does it take for sheets to be laundered and returned to the hotel?', we can say 'From as little as 19 hours to as much as 135 hours'. Certainly there is considerable variation here and it is difficult to see which values occur more often; some organisation is needed.

Counting up the number of each value that occurs in the table above produces the following table of frequencies.

Time taken (hours)	19	24	28	45	47	53	58	63	64	65	67	68	69	70
Frequency	2	1	1	1	1	1	1	2	4	1	1	2	4	3
Time taken (hours)	71	72	74	76	78	79	80	81	82	83	84	86	87	88
Frequency	2	2	2	1	1	2	2	2	4	1	3	3	2	1
Time taken (hours)	89	93	94	95	96	99	102	105	110	111	113	115	116	119
Frequency	3	2	3	2	1	1	1	1	2	1	1	1	1	2
Time taken (hours)	120	123	125	126	127	129	130	135	138					
Frequency	1	1	1	1	2	1	1	2	1					

The frequency table is hardly more useful to look at than the original figures because few values are repeated exactly. Several are very close to one another, and another way of organising the data must be found which incorporates this idea of the 'closeness' of values. One way is to count how many observations fall into a class or category.

Frequency table – times to process 86 batches of sheets

Range of values of time in hours (class)	1–20	21–40	41–60	61–80	81–100	101–120	121–140
Frequency	2	2	4	29	28	11	10

This table performs a similar function to the frequency table on page 12 and from it a diagram may also be drawn. This is called a histogram. (Strictly speaking the histogram would have 'frequency density' and not 'frequency' on the vertical scale. Ask your lecturer for the reason.)

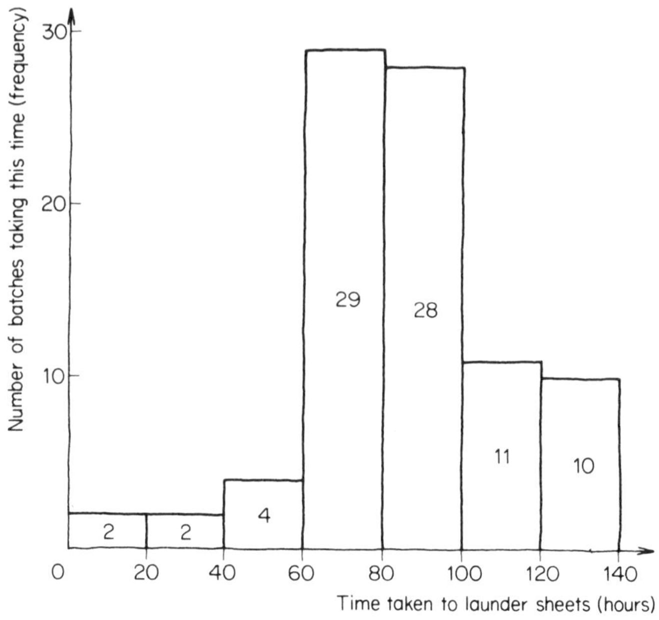

Each block of the histogram can be thought of as actually containing the data in that class. The complete diagram contains the 86 batches.

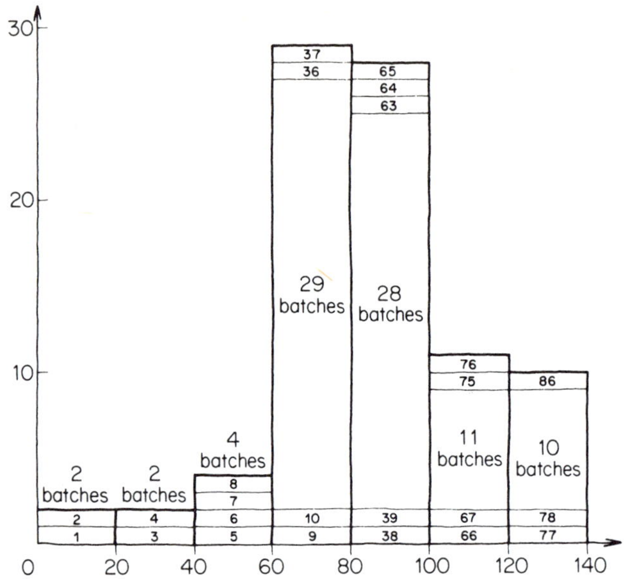

By putting the data into classes the pattern of behaviour has become very clear. Although some batches may be laundered in a very short or a very long time, most take between 60 and 100 hours. The reason for collecting the data was to determine whether the sheets could be processed in less than 4 days. If this were so, then for guests staying for more than 4 nights (long-stayers) only 2 pairs of sheets would be needed: one pair on the bed, the other pair being washed.

From the data we have, 65 batches of sheets were processed in under 100 hours (approximately 4 days). This represents 65/86 × 100, i.e. about 75 per cent or three-quarters of the total. Therefore, for three-quarters of all occasions, 2 pairs of sheets are enough for the long-stayer. The other quarter of the time, sheets take rather longer to process but not longer than 6 days according to the data we have. One extra pair is needed for when the sheets need changing and the spare pair is still at the laundry. So a quarter of the time 3 pairs of sheets are needed: one pair on the bed, one clean pair in the linen cupboard ready to go on to the bed and a third pair at the laundry.

An attempt at an answer to the question 'How many sheets are needed?' can now be made. The number required seems to be 2 or 3 pairs per guest for 'long-stayers', dependent upon how fast the laundry is working. For those who stay for 4 or less nights only 1 pair of sheets is required throughout the visit.

These 'short-stayers' put a heavy burden on the system. If you consider a 7-day period, then 1 guest staying for 7 days requires 2 (or 3) pairs of sheets, but 7 guests each staying 1 day, one after the other, require 7 pairs.

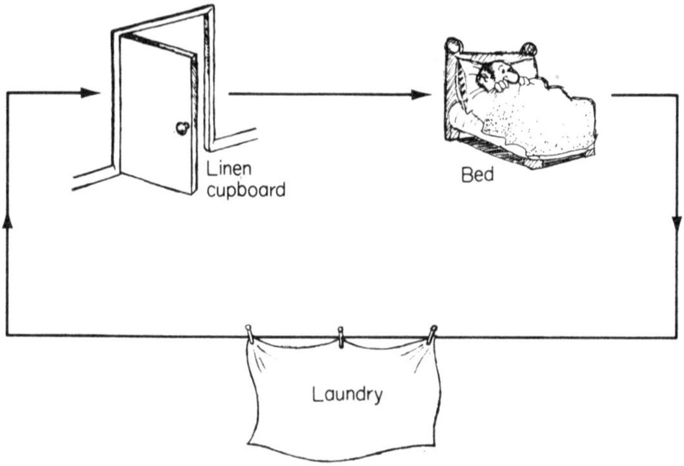

The total linen requirements for one day's arrivals at the hotel depends on how long guests stay. This information is readily available from the data in the table on page 12. From the table, 54, just over half of the 90 guests who arrive on Saturday, 22 June are short-stayers and the other 36 are long-stayers. So the total pairs of sheets needed for the 90 guests during their stay

$$= 54 \text{ @ } 1 \text{ pair} + 36 \text{ @ } 2 \text{ (or } 3\text{) pairs}$$
$$= 54 \times \boxed{1} + 36 \times \boxed{2} = 54 + 72 = 126$$
$$\text{or} = 54 \times \boxed{1} + 36 \times \boxed{3} = 54 + 108 = 162$$

An optimistic answer, based on a fast laundry service would be 126; a pessimistic answer would be 162. Since the laundry service is sometimes fast and sometimes not, the actual requirement is somewhere in between.

So far, only the 90 guests arriving on a particular day have been provided for, but these are not the only ones staying at the hotel on the night of Saturday, 22 June. Guests who arrived on the 21st, 20th, 19th, 18th June etc. may also be there, depending on the length of their visits, and they have linen requirements too. In fact, we have to consider the guests who arrive EVERY day.

If 90 people arrive every day and behave like the 22 June arrivals then their linen needs could also be calculated. However, there is no reason to believe that 22 June was a typical day or even a typical SATURday.

Different numbers of guests with different patterns of behaviour may arrive on weekdays and on different Saturdays of the year. The following frequency table represents the lengths of visit of 85 guests who arrived on Monday, 24 June.

Frequency table – Monday, 24 June

Length of visit	1	2	3	4	5	6	7	8	9	10	11	12	13	14
Number of guests (frequency)	31	21	9	19	3	0	0	0	0	2	0	0	0	0

About the same number of guests arrive as on the Saturday but their behaviour is quite different: over one-third of the guests stay for just 1 night and almost everyone is a short-stayer. The bar charts, drawn to the same scale, highlight this difference.

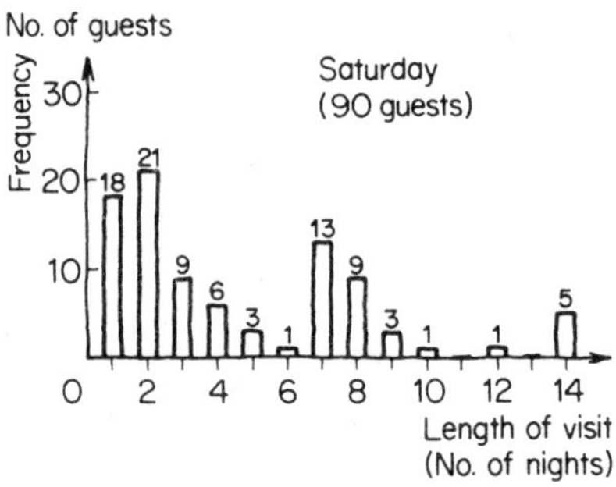

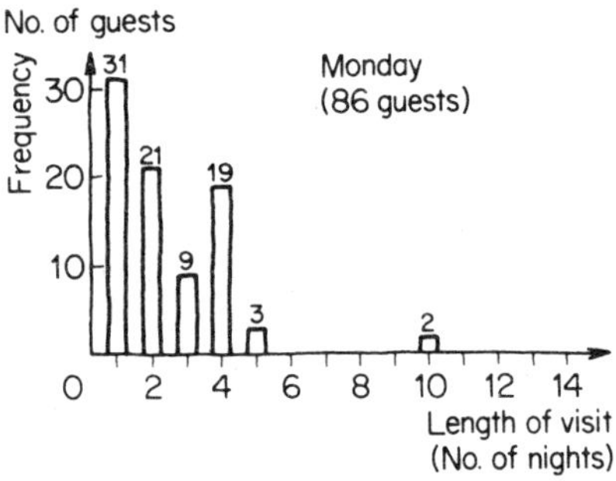

It is clear that Monday's behaviour is not represented by the Saturday pattern. What about the arrivals on other days of the week? The bar charts for these are on the page opposite.

Take a good look at these diagrams. They show very well how the behaviour every day is different. Generally, the guests arriving on Monday, Tuesday and Thursday are short-stayers whereas the arrivals on other days include some long-stayers. Tuesday's and Thursday's behaviour look quite similar and the Sunday pattern looks like a scaled down version of Saturday's pattern. How can we determine how alike or how different the behaviour patterns are? Some more precise level of interpretation is needed to distinguish and compare them.

Interpreting with Measures

How can measures help in describing data? Our everyday descriptions often include measures. Length, thickness and area, for example, are common measures of the shape of an object. Some, like

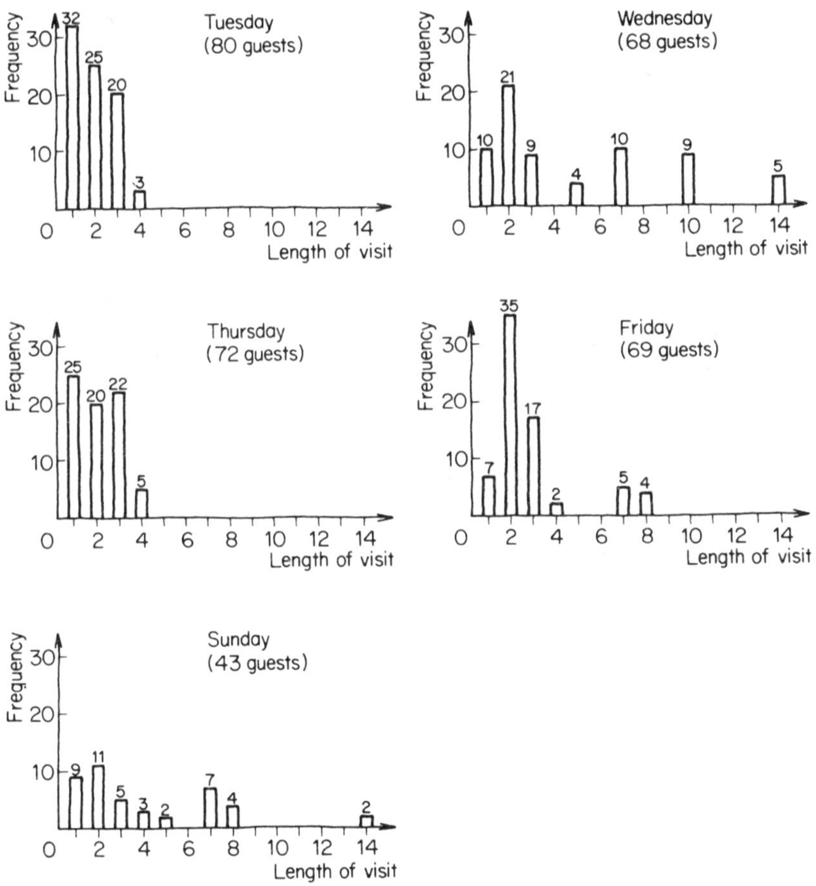

length and temperature, can be obtained by measurement using special tools like the ruler and thermometer; others, like area and volume, rely on calculations and measurement (though tools have been designed to measure these directly).

Each measure provides different but valuable information about the shape and size of the object. The choice of measure used would depend on the reason for describing the shape. Sometimes the description is needed for its own sake so that you might know how the object looks; sometimes it is needed in order to compare it with similar objects. For example, if you wanted to indicate the warmth of a fabric, the thickness may be quoted; you might quote the weight of a tent to show how light and easy it is to carry, while the height and width would be used to show how roomy it is. These measures describe the products and provide a basis for comparison with other rival products. In each case only the appropriate measures are used; others, for example the colour of the tent, are irrelevant when describing the tent's capacity and are best omitted from the description, even when they are available, as they would only confuse the description.

The same practice is followed in Statistics. The measures are called statistics (small s) and are calculated from the data. Each one describes some facet of the behaviour of the variable under study. This enables us to compare one set of data with another set, or with our preconceived notions about how such data should look.

Considering again the hotel data, how can the difference between each day's behaviour be expressed? From the data it seems that guests arriving on some days stay noticeably longer than those arriving on other days. Compare Saturday and Monday for example. A statistic is needed to show up this difference and presumably its value for the Saturday data would be larger than for the Monday data. One such measure is the arithmetic mean: to calculate the arithmetic mean for Saturday's data, the lengths of visits of the 90 guests are added together and the total divided by 90

$$\begin{aligned}\text{mean length of Saturday arrivals} &= \frac{\text{sum of all the values in table (page 11)}}{90} \text{ nights} \\ &= \frac{1+3+4+6+9+\ldots+2+5+2+1}{90} \\ &= \frac{414}{90} = 4\cdot60 \text{ nights}\end{aligned}$$

The 414 is the total number of nights spent at the hotel by the 90 guests; dividing by 90 'shares out' this total equally among these 90. We say that 'on average' each guest stayed 4·60 nights, although no guest actually stayed for 4·60 nights. (In fact it is impossible for someone to stay for a fraction of a night.) Some people stayed for longer and some for shorter periods, but the mean gives some idea of how long guests stay.

Calculating the arithmetic mean for the Monday arrivals we get

$$\text{mean length of Monday arrivals} = \frac{211}{85} = 2\cdot48 \text{ nights}$$

Now we have what we are looking for, two figures that enable us to compare Saturday and Monday. The mean has presented in numbers what the diagrams failed to convey clearly.

The mean of any set of values of a variable can be calculated following the same procedure. Since the values in any set of data vary, it is common practice to call a single value X. Then the mean

$$= \frac{\text{sum of the } X\text{'s}}{\text{total number of } X\text{'s}}$$

Since mathematicians are possessed by evil symbols there is one for 'the sum of'. The symbol is Σ, the Greek letter 'sigma'. The total number of X's is often represented by N, and the mean by $\bar{X}$. So the formula for the mean looks like this:

$$\bar{X} = \frac{\Sigma X}{N}$$

Read it like this: 'X bar equals sigma X divided by N'.

Practise the calculation by finding the mean length of visit for the guests who arrived on other days and check your answers with the table below which contains the mean length of visit for each day's arrivals. Study the table carefully.

Start of visit	*Mean number of nights stayed*
Sunday	4·12
Monday	2·48
Tuesday	1·92
Wednesday	4·84
Thursday	2·10
Friday	2·94
Saturday	4·60

This table confirms that Sunday IS very like Saturday (even though less people arrived on Sunday); Wednesday is also like these two. What is it about Wednesday that makes it like Saturday or Sunday? It may be that Wednesday's arrivals contain a contingent of long-stayers travelling in the comfort of midweek.

The other days are more like each other, Tuesday having the shortest visits and Friday the longest. The latter is probably due to weekenders.

So much for the comparison of means, but what does each mean tell us about its own set of data? It is supposed to indicate 'typical' behaviour. If Mr Average-Tuesday arrival came to the hotel we would expect him to stay about 1·92 days. Well, would we? Looking at the diagram for Tuesday, most people stay 1 or 2 days so the mean, 1·92, is not a bad measure. We cannot say the same for Saturday's mean, 4·60 nights. Hardly anyone actually stays 4 or 5 nights: 9 guests in fact out of the 90; people either have short stays of 1 or 2 days or they stay for 7 or 8 days. This statement accounts for 61 of the guests, about two-thirds of the total and so is more useful than quoting the mean. This idea of quoting popular values has led to the use of a statistic called the mode, the most frequently occurring (fashionable) value. A set of data may have several modes, shown as peaks on the histogram, or bar chart.

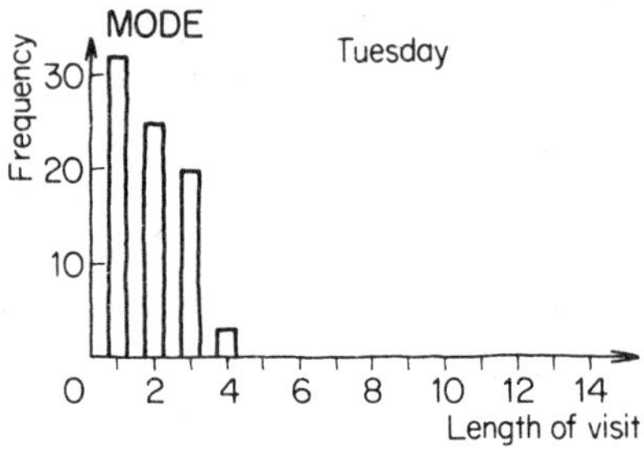

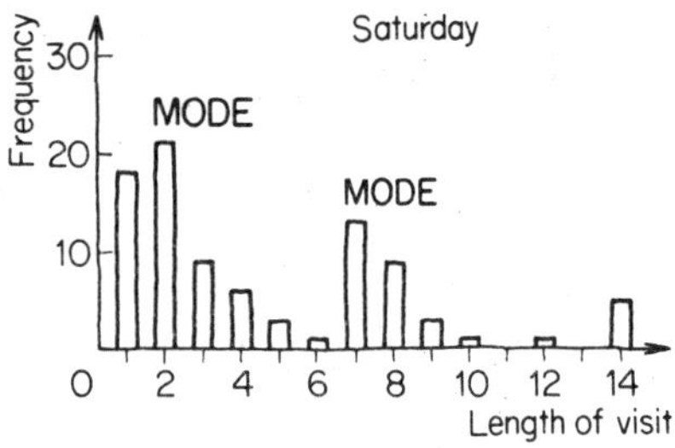

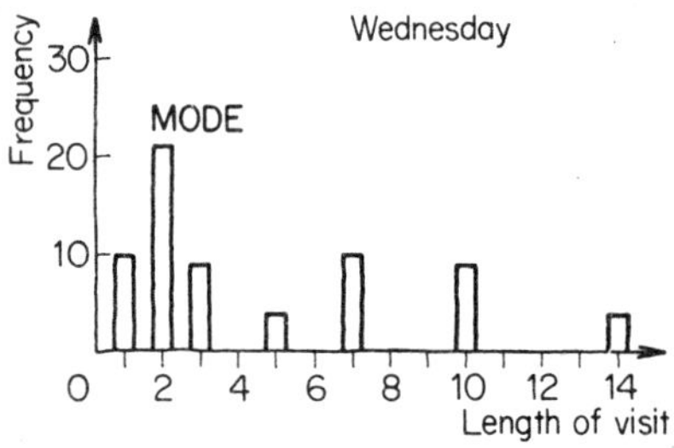

Tuesday has one mode of 1 night. Saturday has 2 modes of 2 and 7 nights; more people stayed for 2 nights than for any other period. Wednesday has a mode of 2 nights which is more indicative of typical Wednesday behaviour than the mean, which is 4·84. The trouble with the mean is that it is sometimes too sensitive: if some crank decided to stay for 1 year, for instance, and arrived on a Wednesday, the mean would become artificially high at 10·06 nights. The mean has shared out all these extra nights evenly between the guests, but this distorts the real picture.

This has happened with Saturday's data; a general pooling has taken place producing a value for the mean that represents nobody. The mean IS the appropriate measure when the data contains a few high and a few low values, but plenty in the middle. This is the case for Monday, Tuesday, Thursday and Friday, the bar charts forming a 'humped' pattern. Where the data forms a 'U' pattern (e.g. Saturday's), the mode (or modes) is more appropriate.

Using the mean or the mode (and your lecturer will tell you of other statistics which can do a similar job), some idea of an average or typical length of visit can be calculated for each day. However, another characteristic of the diagrams above is that the lengths of stay for a particular day are similar on some days and dissimilar on others. Monday's visitors, for example, are more like each other in their behaviour than are Saturday's; the latter display more varied behaviour. The mean is not designed to show this so we have to consider another type of statistic, which measures variation. If the visitors behaved like sheep, all staying for the same period, then there is no variation and the statistic should have the value of zero; if they all stayed for different periods then the value should be large.

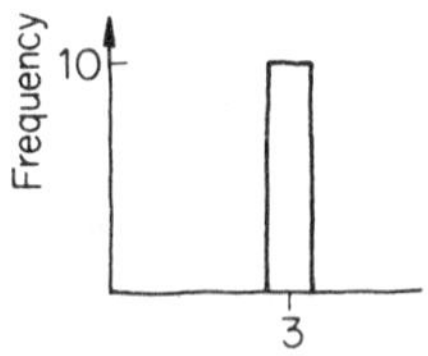

(i) Sheep

(ii) Individuals

An obvious choice is to quote the smallest and largest visit. This is called the range.

Day of arrival	*Range (number of nights)*
Sunday	[1, 14]
Monday	[1, 10]
Tuesday	[1, 4]
Wednesday	[1, 14]
Thursday	[1, 4]
Friday	[1, 8]
Saturday	[1, 14]

The ranges for Tuesday and Thursday, when used together with the mean, give a good description of the data. But the range is sensitive to very low or high values: for example, Monday's range implies more variation in the data than there really is.

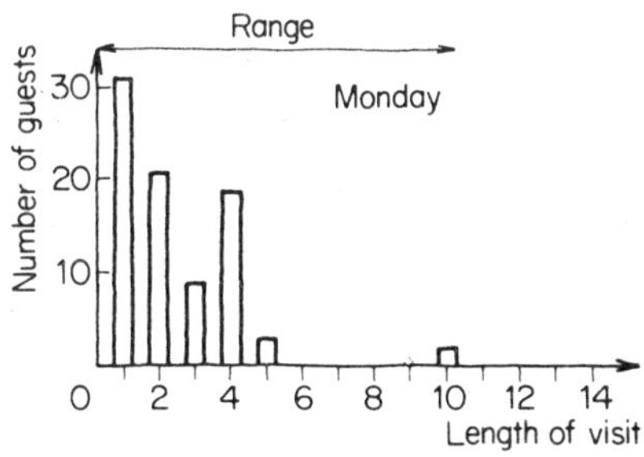

Another criticism is that no account is taken of the values between the endpoints of the range. The data can 'hump' together or it can spread itself all over the place and the range will have the same value, e.g. compare Sunday and Wednesday.

(i) 'Hump' together

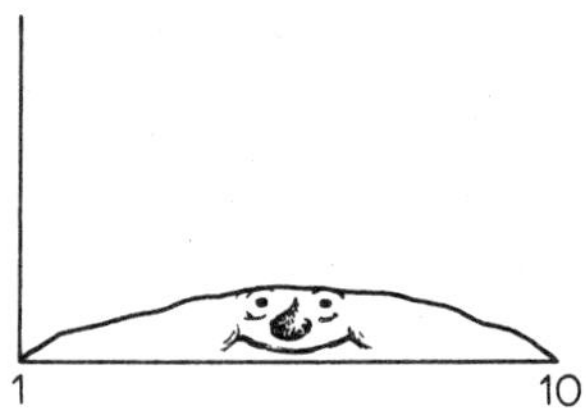

(ii) 'Spread itself all over the place'

A measure of variation which is more discriminating than the range is the standard deviation. This tells us how far a typical observation would be from the mean. The calculation of this statistic involves the subtraction of the mean from every observation to find out how far away from the mean each one is. To illustrate, consider the first ten of Saturday's arrivals:

1, 3, 4, 6, 9, 1, 2, 1, 1, 2

STEP 1: First the mean is calculated.

$$\text{Mean} = \frac{\Sigma X}{10} = \frac{30}{10} = 3 \text{ nights}$$

STEP 2: Subtract the mean from each value.

X	1	3	4	6	9	1	2	1	1	2
$d = X - \bar{X}$	−2	0	1	3	6	−2	−1	−2	−2	−1

This tells us how far each observation is from the mean. Clearly, some are farther away than others; the difference is called d.

STEP 3: This step seems rather peculiar. Instead of finding the mean of the differences, $\bar{d}$, we square them and find the mean of the SQUARED d's. [Calculate $\bar{d}$ yourself and see what you get. Calculate $\bar{d}$ for another 10 values of X. Ask your lecturer to explain why you get this value.]

$d^2 = (X - \bar{X})^2$	4	0	1	9	36	4	1	4	4	1

$$\text{Mean of the } d^2\text{'s} = \frac{\Sigma d^2}{10} = \frac{37}{10} = 3 \cdot 7$$

So a typical 'squared deviation' is 3·7 square nights.

STEP 4: We don't want to know anything about square nights so in this step we take the square root

$$\sqrt{\frac{\Sigma d^2}{10}} = \sqrt{3 \cdot 7} = \text{a bit less than } 2$$

This is the average deviation from the mean and is called the STANDARD DEVIATION. The general formula is

$$\boxed{\text{S.D.} = \sqrt{\frac{\Sigma(X-\bar{X})^2}{N}}}$$

This formula can be put into a different form which is useful when the mean is an awkward value (as is usually the case):

$$\text{S.D.} = \sqrt{\frac{\Sigma X^2}{N} - \left(\frac{\Sigma X}{N}\right)^2}$$

Ask your lecturer to demonstrate that this is the same formula as the first one. Notice the difference between 'sigma, X – squared' and 'sigma – X, squared' in the formula.

Various coding methods are available to ease hand calculation of the mean and standard deviation and it is sometimes worthwhile employing them if a calculating machine is not available. These are illustrated in Chapter 10.

The standard deviations for each day's arrivals are as follows:

Day of arrival	*Standard deviation (number of nights)*
Sunday	3·27
Monday	1·72
Tuesday	0·89
Wednesday	3·94
Thursday	0·96
Friday	1·85
Saturday	3·60

This statistic sorts the data into two categories: the guests whose behaviour varies little: Monday, Tuesday, Thursday, Friday; and those whose behaviour varies a lot: Sunday, Wednesday and Saturday arrivals.

Day of arrival	*Mean*	*Standard deviation*	
Monday	2·48	1·72	
Tuesday	1·92	0·89	
Thursday	2·10	0·96	MTTF type
Friday	2·94	1·85	
Sunday	4·12	3·27	
Wednesday	4·84	3·94	SWS type
Saturday	4·60	3·60	

There seems to be a relationship between the day of arrival of a guest and his length of visit behaviour. Guests arriving on Mondays, for example, stay for short periods and behave very much like each other (indicated by the standard deviation of only 1·72 nights). Guests arriving on Sunday, however, stay for longer periods on the whole, but they vary much more among themselves, some staying for short and some for long periods (indicated by the larger value of the standard deviation).

Moreover, there seem to be only two behavioural types: the MTTF and the SWS. Each day seems to belong to one of these categories as shown in the previous table. The question now arises, 'Is the behaviour of guests for each day of a category essentially the same?' Are Monday's guests like Tuesdays? Are Saturday's like Sunday's? Is it possible to extract one behavioural type to describe the MTTF guests and one for the SWS guests? The answer is 'yes' but not from this level of interpretation.

Interpreting with Percentages

The comparison of the behaviour patterns represented by the frequencies is confused by the fact that different numbers of people arrive each day, for example 43 on Sunday and 90 on Saturday. Somehow, to compare the patterns of behaviour of each day in the MTTF category, the effect of different numbers of people arriving must be removed. The correct statistic for comparing the patterns is the proportion or percentage of the total number of guests who stayed a certain length of time.

For example, on Monday, 31 out of 86 guests stayed for 1 night. This represents 31/86 of the total, a proportion of 0·36 or a percentage of 36. Using percentages the patterns for different days can be compared.

Percentages of guests staying for different lengths of visit. Each row applies to those guests who arrived on a particular day

	Length of visit	1	2	3	4	5	6	7	8	9	10	11	12	13	14
	Monday	36	24	10	22	3	0	0	0	0	2	0	0	0	0
A	Tuesday	40	31	25	4	0	0	0	0	0	0	0	0	0	0
	Thursday	35	28	31	7	0	0	0	0	0	0	0	0	0	0
	Friday	10	51	25	3	0	0	7	6	0	0	0	0	0	0
	Sunday	21	26	12	7	5	0	16	9	0	0	0	0	0	5
B	Wednesday	15	31	13	0	6	0	15	0	0	13	0	0	0	7
	Saturday	20	23	10	7	3	1	14	10	3	1	0	1	0	6

Study the tables carefully. You can see from table A that the patterns for the four days are similar, by noting where the zeros and non-zeros appear.

Monday	×	×	×	×	×	—	—	—	—	—	—	—	—	—
Tuesday	×	×	×	×	×	—	—	—	—	×	—	—	—	—
Thursday	×	×	×	×	—	—	—	—	—	—	—	—	—	—
Friday	×	×	×	×	—	—	×	×	—	—	—	—	—	—

However, a comparison of the VALUES of the proportions shows that Friday's behaviour is different from the other three days. This is a piece of information that could not be clearly seen from the other three levels of interpretation, although the bar charts of the organised data hinted at it. Look at table B and see what you can say about the behaviour of Sunday, Wednesday and Saturday arrivals relative to each other.

Modelling

You should now feel you know the behaviour of hotel guests very well – at least the ones who arrived during that week in June. What happens in other weeks is a different story and by now you should have realised that to find that out, data needs to be collected for arrivals in all the other weeks in the year. You might find two kinds of behaviour, a winter and a summer behaviour, in much the same way as we found two types of days. Then behaviour patterns for summer weeks would look alike and in particular all MONdays in summer would look alike. This would greatly simplify the linen problem for the hotel manager for he could then design his linen system so that it copes with the two types of behaviour: summer and winter. It would be possible to extract seven patterns, one for each day in summer, and another seven winter patterns. For example, a SUMMER-MONDAY pattern would be obtained by compiling tables of the form of tables A and B for each week of summer and averaging the percentages. An example is given where three weeks' data are available.

Length of visit	1	2	3	4	5	6	7	8	9	10
Monday, 24 June	36	24	10	22	3	0	0	0	0	2
Monday, 1 July	34	26	15	17	2	1	0	0	1	2
Monday, 8 July	32	26	13	19	4	1	1	0	0	2
Average percentage	34	25	13	19	3	1	0	0	0	2

The pattern made up by the average percentages would be taken as typical of summer-Monday behaviour. This pattern would be called a model, a Summer-Monday model, because it models the behaviour of a particular kind of guest.

Models are extremely useful in interpreting data because they help to simplify the practical situation. One model often applies to several situations, as we have just seen. This is not to say that every aspect of each situation is incorporated in the model; only those considered essential will be considered. Many factors affect length of visit. But the purpose of developing a length of visit model is to investigate linen requirements. Clearly, factors such as the place of residence of the guest are irrelevant, and will be omitted from the model; other factors like the purpose of the visit will also be considered irrelevant, but this assumption may be relaxed in different circumstances (e.g. if there is a football match as on 19 June).

Probability Models

The summer-Monday model is an example of a model derived directly from the data. Often it is possible to construct a model of a situation BEFORE any data is collected. For example, before you toss a coin 20 times it is possible to say what percentage of the results is likely to be heads and what percentage tails. The answer is 50 per cent in each case, provided the coin is unbiased. This is because you know from the nature of the coin that there is an equal chance of a head or a tail at each toss. We can measure chance. The measure is called probability, and it has values ranging from 0 (no chance!) to 1 (absolute certainty). The probability of obtaining a head is $\frac{1}{2}$: this is a measure of our certainty, before the coin is tossed, that the result will be a head.

Specifying the probabilities in the coin situation is easy. It relies on the assumption of the unbiasedness of the coin which is based on our knowledge of its shape. A real-life situation is more complicated. It is difficult to describe the mechanics of a real-life situation before data is collected on it, therefore the probabilities are often not known. In these cases the proportions can be used as estimates. For example, if the assumption of unbiasedness could not be made, the coin could be tossed several times and the number of heads counted. If 11 heads are obtained when the coin is tossed 20 times then the proportion of heads is 11/20 (= 0·55) and this is an estimate of the probability of getting a head. If the coin were tossed 40 times and 18 heads resulted, the proportion is 18/40 (= 0·45). Hopefully, the more data we collect (by tossing the coin more times), the nearer the proportion should be to the true probability, so the better it is as an estimate.

Number of tosses	*Number of heads*	*Proportion*
0	0	0·00
20	11	0·55
40	18	0·45
60	29	0·48
80	41	0·51
100	49	0·49

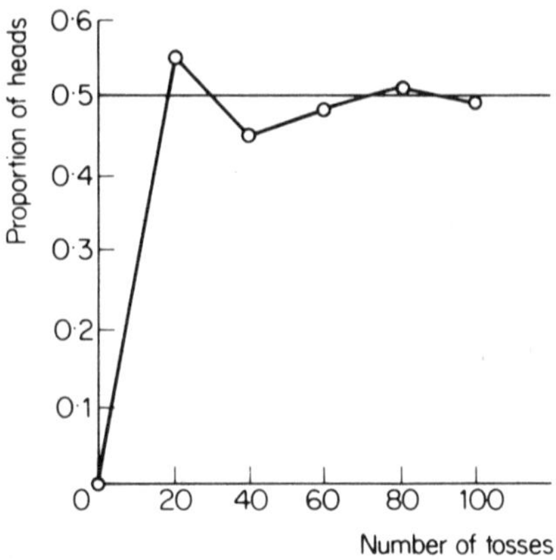

Probabilities can be determined for the hotel data in the same way and the model can be expressed in terms of these rather than the percentages.

Summer-Monday probability model

Length of visit	1	2	3	4	5	6	7	8	9	10
Probabilities	0·35	0·26	0·13	0·20	0·03	0·01	0	0	0	0·02

Some probability patterns occur very often. The histograms representing the data to which the pattern applies look very similar; this means that one model fits many practical situations. One such model is the Normal model. The probabilities are called the Normal probabilities and the following graph is an abstraction of the shape of the histogram.

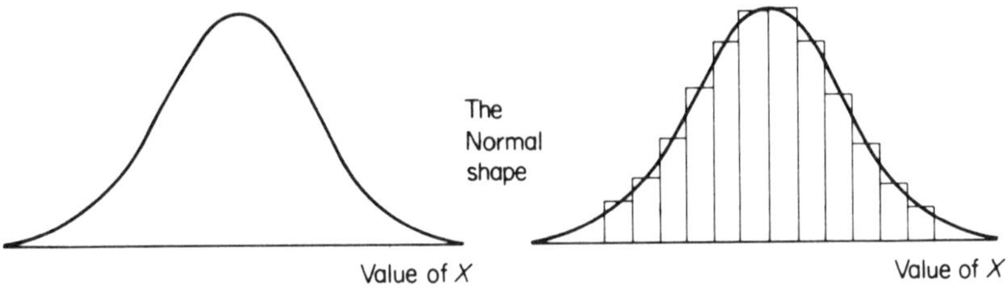

Middling values of *X* are the most popular, and extremely small or large values the least popular. The shape is symmetrical about the 'middle' value which is in fact the mean value of *X*. An important feature of the Normal model is the arrangement of the percentages of the values of *X* occurring. The Normal model is called the 'Normal distribution' because it specifies how the percentages or probabilities are distributed among the values of *X*.

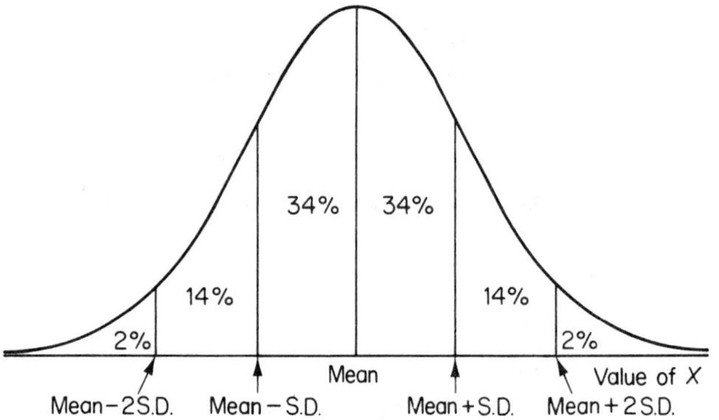

The distribution of probabilities in the Normal model

We have already encountered a histogram of this shape, the diagram of the time to process 86 batches of linen. The mean length of time for this data is about 85 days and the standard deviation is 25 days. The histogram should therefore be symmetrical about 85 and the percentages of data in each category should be as follows, if the length of time to process linen has the Normal pattern.

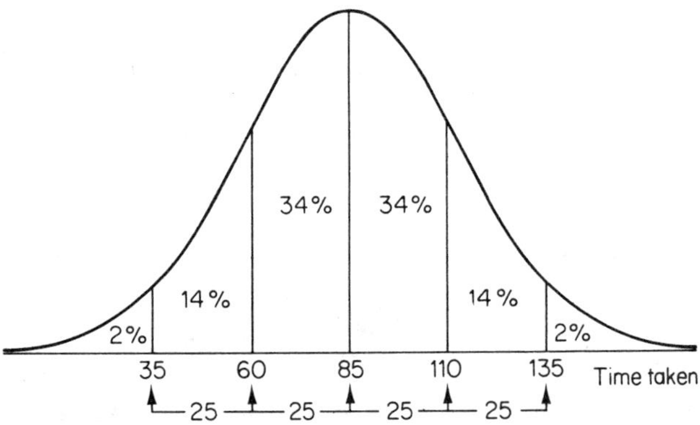

Theoretical behaviour – linen times

The percentages from the data don't exactly match, but the 'fit' of the Normal model is not too bad. The data does have 68 per cent of its values in the middle two blocks.

The histogram is based on only 85 observations; if more data were collected the 'fit' might be better. Tests are available to determine whether a particular model like the Normal model can be used to represent the behaviour of a variable.

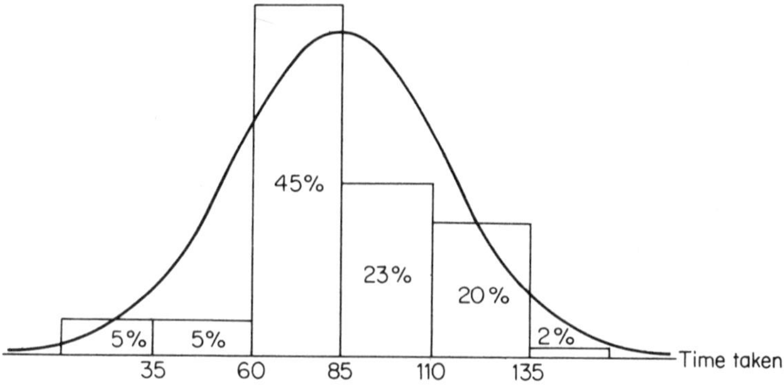

Actual behaviour – linen times

There are certain situations for which we know the Normal model applies, even before data is collected. Just as is the case with the coin, our knowledge of behaviour allows us to specify the model and hence the probabilities. If an article is being produced to some specification, length, diameter, weight etc., then the production process aims to meet this, but sometimes the article may be too small and sometimes too large. Most of the time it should be equal or very close to the specification.

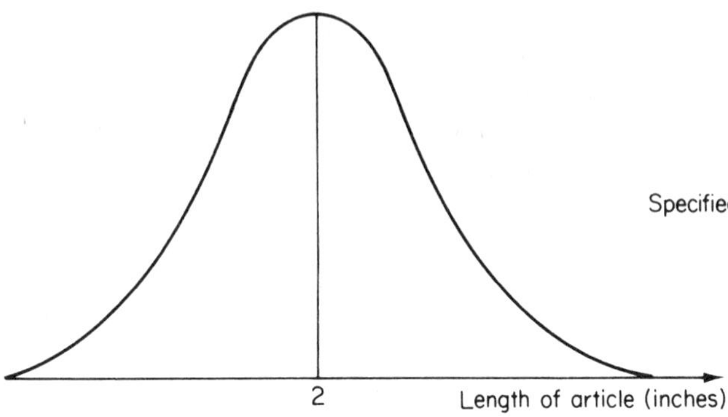

A Normal model

This is one case where a Normal model would apply. Of course many models could be used, depending on the standard deviation of the length.

The model in figure (*a*) would apply to a fairly precise production process; very short or very long articles would not be made very often. The model in figure (*c*) would apply to a rather sloppy process. In order to find the appropriate model to use, some data needs to be collected to estimate the standard deviation of the variable. The

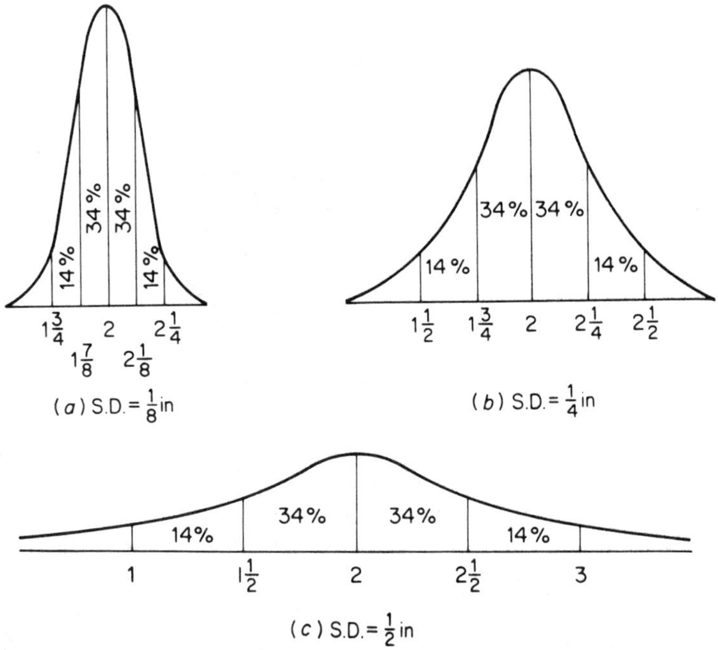

Examples of different Normal models

advantage of knowing that the Normal model can be used is still retained, because the amount of data needed to estimate a standard deviation (15 or 20 values) is still less than that needed to establish the shape of a histogram (50 to 100 values).

Another practical situation to which the Normal model applies is one which behaves AS IF it were trying to produce something to a specification. Dog breeding is an example. The parents may be of several sizes. If litters were produced without control for many generations the very large and very small dogs would be rare; a 'cancelling out' of these would take place and most dogs would be of a middling size. The weight of dogs, for example, would have a Normal distribution. The breeding process behaves as if it were trying to produce an 'average dog' with a certain weight; this mean weight, as well as the standard deviation of weight, would have to be estimated from the data in order to decide which Normal model to use.

There are many situations like the dog-breeding case where 'cancelling out' of extreme values occurs. The variable under investigation is often made up of, or dependent upon, several factors each of which also varies and the result is a variable with a Normal 'shape'. Bearing this in mind it is not surprising that the time taken to process linen has an approximate Normal shape: it is made up of the times for collection, machine loading, washing, unloading, drying, ironing, packing and delivery. However, just because the Normal model is often applicable (and we wouldn't bother to have it as a model unless this were so) it does not apply to every situation, for example the length of visit behaviour of hotel guests; in these circumstances some other model must be used.

Monitoring the System

When data has been collected and behaviour is more fully understood, perhaps being represented by a model, only then can a system be designed to cope with this behaviour. The system currently in operation may as a result be altered or completely replaced. When the new system has been installed, data collection still continues. A system cannot be allowed to carry on without observation if it is to work well: the behaviour of guests may change, they may generally stay for longer periods and their linen needs would change. The system which was designed for one kind of behaviour might not cope with this new kind, it may over-provide or be totally inadequate. Instead of waiting for the resulting difficulties to arise it is preferable to monitor behaviour by collecting data on it and so detect any changes as soon as they occur and adjust the system accordingly. The change might still cause a problem, but knowing about it in good time provides the opportunity to deal with it before it grows into a bigger problem. Sometimes the system may not be working properly: there is some break-down or delay which may go unnoticed until a crisis occurs, and once again the parts of the system need to be monitored by collecting information on them. Each week, for example, a histogram of the times taken to process linen could be compiled and this could then be compared with the model of process times based on the system working properly, either to confirm that all is well or to highlight a problem.

Business is concerned with systems. Data is collected for the design, monitoring and change of systems. It is the business of the businessman to be able to read what the data says.

3

Line Up the Function

In Chapter 2, graphical methods were used to describe the properties of a variable. The time taken to process a batch of linen through the laundry would vary from batch to batch and this VARIATION could be represented on a diagram called a histogram.

Another way of graphically representing data can be used when the question is 'How does the time to process the linen vary depending on the day of the week?'

We want answers to such specific questions as: Is Friday's time longer than Monday's time? If so, by how much? Does the time increase regularly throughout the week, or is the pattern irregular? What is the pattern?

Patterns

Let us look at the two sets of figures.

		Time (mins)
Monday	(Day 1)	100
Tuesday	(Day 2)	200
Wednesday	(Day 3)	300
Thursday	(Day 4)	400
Friday	(Day 5)	500

		Time (mins)
Monday	(Day 1)	500
Tuesday	(Day 2)	300
Wednesday	(Day 3)	100
Thursday	(Day 4)	400
Friday	(Day 5)	200

In the first case there is an obvious pattern: each day the time taken increased by a fixed amount – 100 minutes. The rate of increase in time taken is 100 minutes/day.

The pattern in the second list is very difficult to detect. When displayed graphically the pattern or lack of it should stand out even more clearly.

Since the points in (*a*) follow a distinct pattern, they can be replaced by the pattern itself, a straight line as in (*c*).

How are we supposed to read diagrams like (*c*)? What are we looking for? What information is there in the diagram for us to find?

The general method of attack is as follows:

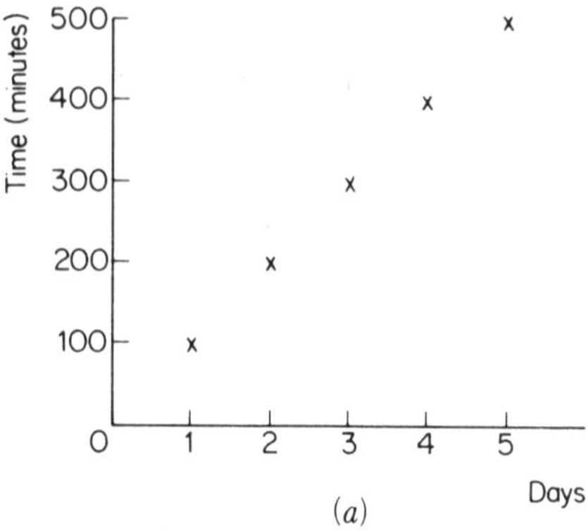

(*a*)
Points fall on a straight line

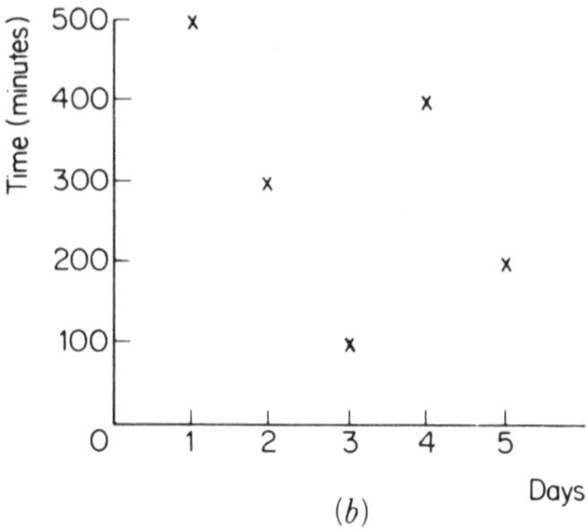

(*b*)
No obvious pattern

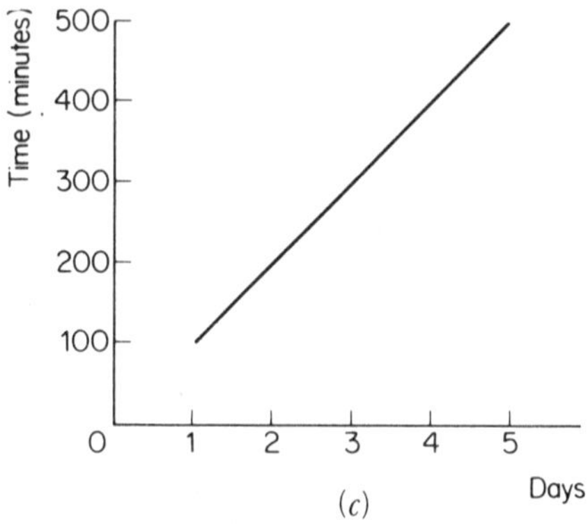

(*c*)
The straight line model for the points in (*a*)

(1) Look at the axes (the vertical and horizontal lines with numbers on) to find out two things:

(*a*) the names attached to them;
(*b*) the scale recorded on each axis.

(2) Decide which axis name 'depends' on the other; this is usually attached to the horizontal but you cannot always depend on this.

(3) What sort of pattern is there? A straight line? A curve? A SERIES of curves?

(4) Reading the graph from left to right we ask what are the important quantities which define the pattern. Does the pattern start with a high value on the vertical axis or does it start low down?

Does it increase as we move left to right or decrease (if no one is looking trace the line in (*c*) with your finger from left to right to get the idea of a pattern which increases from left to right).

If it does increase, how fast does it increase?

We need a scale for measuring the speed or rate of increase

which has more points on it than just fast and slow. Without such a measurement, graphs can be very misleading because the SAME information can be displayed in apparently quite different ways.

Compare the following graphs by following the method of attack (1)–(4), making notes of your answers.

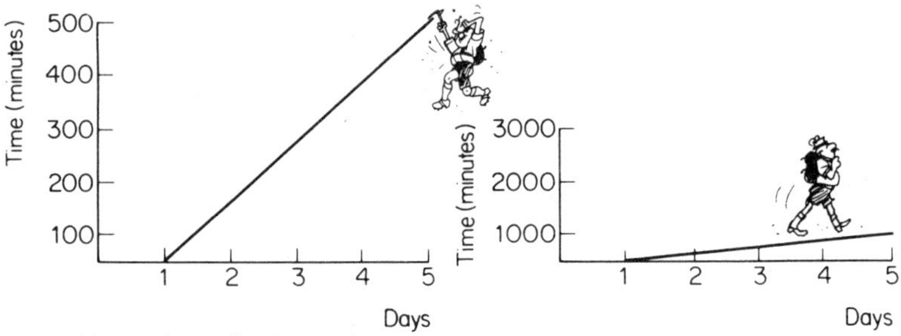

My goodness that's steep! *Just a gentle slope!*

This idea of drawing a straight line to replace, and describe, a set of values is used extensively in business, science and social enquiry. Different disciplines use the idea in slightly different ways so a thorough understanding of the straight-line model must be developed. We can start by considering a mythical company.

Suppose a company has profits of £3 million, £4 million and £5 million in the years 1970–2 and suppose you were asked to predict the sales for 1973 given no further information about markets, world conditions, interest rates and other such vital matters. Then you would probably predict, all other things being equal, that the profits would be £6 million. This guess does not qualify you to set up as a management consultant in forecasting but it does show that you have an initiative understanding of a straight-line model, the characteristic of which is a constant increase in the profits each year.

Plotting the data conforming to the straight-line model, we have the diagram below:

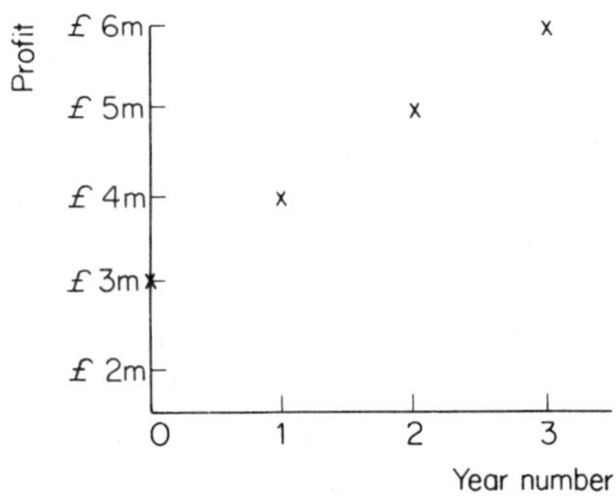

and the model that the data suggests in the following diagram.

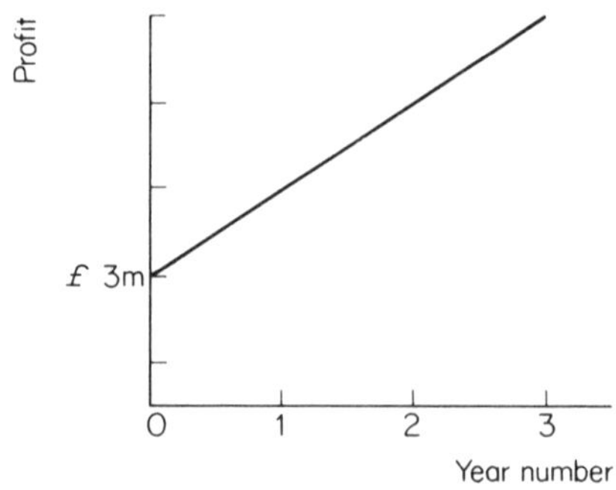

In the year 0 the profit is £3m and in each succeeding year it increases by £1m.

For instance we could write

$$\text{Profit in year } 4 = £3\text{m} + £1\text{m} \times 4.$$

This is the equation relating profit to year number. All that you have to do to calculate the profit expected at a particular time in the future is to use the equation given below. Put the number of the appropriate year in each of the boxes in the equation and work out the arithmetic on the right-hand side. The example uses year number 8.

The equation is,

$$\text{Profit in year } \square = £3\text{m} + £1\text{m} \times \square$$

and it can be used to find profit in any year.

For example,

$$\begin{aligned}\text{Profit in year } \boxed{8} &= £3\text{m} + £1\text{m} \times \boxed{8}\\ &= £3\text{m} + £8\text{m}\\ &= £11\text{m}\end{aligned}$$

You choose the year and calculate the profit!

The model equation can be simplified to read

$$P\,\square = 3 + 1 \cdot \square$$

and even the box can be simplified by saying that what goes in it is a number which could vary and is written as x, where x corresponds to a year number.

The model is now

$$P(x) = 3 + 1x$$

$P(4)$ is read as the value of the profit in year 4.

The relationship between the equation and the graph is seen below.

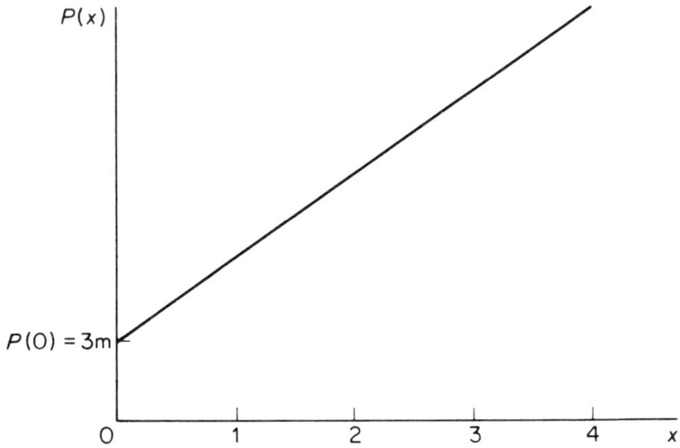

There are two essential features of the straight line: the first is $P(0)$, which is the value of P when x is 0.

$$\text{NOW} \quad P(0) = £3\text{m} + £1\text{m} \,.\, 0$$
$$= £3\text{m}$$

The second is the rate of increase or slope.

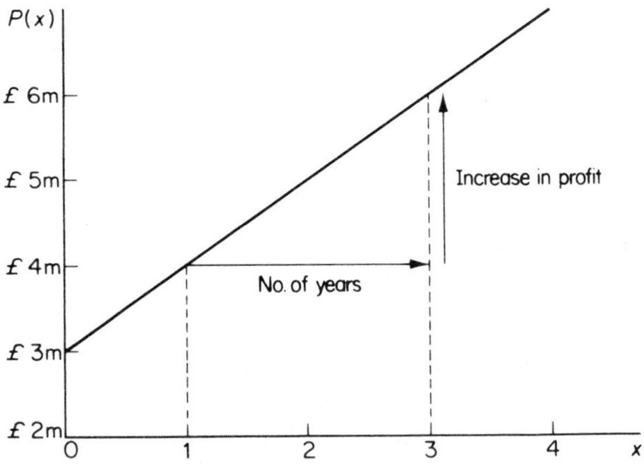

In the two years from $x = 1$ to $x = 3$, profits have increased from £4m to £6m.

$$\text{Slope} = \frac{\text{increase in profit}}{\text{number of years}} = \frac{£2\text{m}}{2} = £1 \text{ per year}$$

This indicates that every year the profit increases by £1m.

We can now rewrite the straight-line model as

$$P(x) = P(0) + \text{slope} \cdot x$$

It is only necessary to know $P(0)$ and the slope to define a line. Different values of $P(0)$ and slope will define different lines.

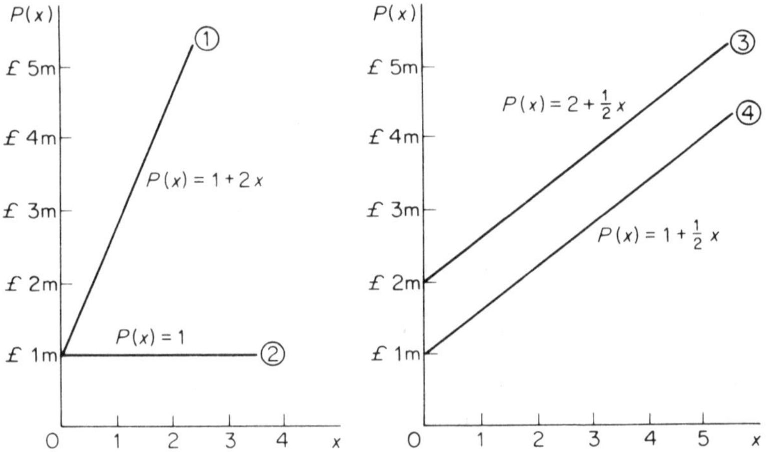

Lines 1 and 2 have the same $P(0)$ value of £1m but different slopes: line 2 has no slope at all so the profit remains the same at £1m for every value of x. Lines 3 and 4 have identical slopes of £$\frac{1}{2}$m/year but they cut the vertical axis at different points so their $P(0)$ values differ.

Not all lines rise from left to right. In the following figure the firm's profits decrease every year by £1m and the graph of the equation

$$P(x) = £5\text{m} + (-£1\text{m}) \cdot x \text{ is}$$

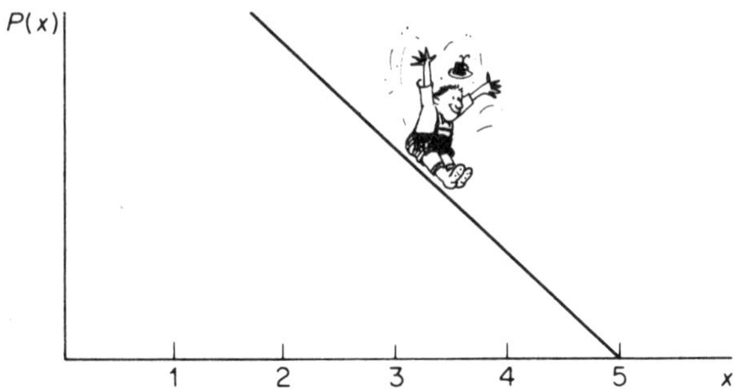

The equation would usually be simplified to

$$P = 5 - x$$

profit being expressed in millions of pounds.

When you are confronted by a straight-line relationship, one of the first things to check is whether there is a negative sign in front of the x term: this indicates that the slope is downwards. Naturally the slope would not be downwards if we were reading the graph from right to left but we adopt the convention that graphs are read from left to right.

If the relationship between two variables can be represented by a straight line then ALL the information about the relationship will be contained in the answers to the following questions:

(*a*) Which variable is the dependent one?

This variable will be plotted vertically like Profit in the example. It is called the DEPENDENT variable because its values we calculated depend on the values of the other variables, the year number in this case.

(*b*) What letters should be chosen to represent the variables?

P and x were used in the example but many books use y and x; y usually denoting the DEPENDENT variable. This is a dull convention and where possible you should allow your imagination free rein in naming variables.

(*c*) What are the values of the slope and intercept?

Now you need to practice the vocabulary we have been developing. There is an exercise to say out loud before breakfast every morning.

Function

Sales *depend on* advertising expenditure
Sales *is a function of* advertising expenditure
Sales = constant + (another constant) times expenditure
Sales = intercept + slope exp.
$S = C + M \, . \, x$

C and M are constants which would be fixed for a particular firm at a particular time; x would be the variable expenditure, and S the sales which would result from that expenditure. Sales is the dependent variable. Sales managers dream of a simple relationship like this between sales and expenditure! The actual relationship is much more complicated and $S = C + Mx$ could only be a very rough model of the real situation. This idea of function is very interesting.

When we say sales is a function of expenditure and write

$$S = S(x)$$

or even $S = f(x)$, we are merely asserting that there is some formula or procedure, which *might* be very complicated, which enables a sales value to be calculated for every value of x, the expenditure figure. The simplest of these procedures is the one embodied in the equation of the straight line: the LINE or function.

When we write

$$S = f(x) = 10 + 8x$$

– remembering, of course, that 10 is $f(0)$ and 8 is the slope – we are merely asserting that when there is NO advertising, sales would still be 10 and that for a unit increase in advertising expenditure there will be an increase of 8 in sales. Quickly plot S against x for values of x from 0 to 10 (you will need to calculate S for $x = 0$, $x = 5$ and $x = 10$ say) and check that when $S = 14$, $x = \frac{1}{2}$ by plotting it on your graph.

Functions in General

Unfortunately, most relationships are not linear and when you see

$$t = g(z)$$

natural caution should encourage you to read it as some variable t (to be plotted vertically) depends in some way on another variable z. The depends-in-some-way is denoted by the g in front of the bracketed z.

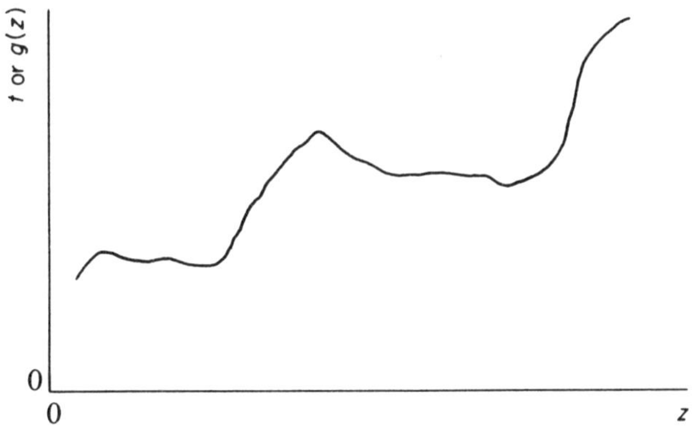

There is no harm in 'hypothesising' that $t = g(z)$ is a linear function and hypothesising that when z is zero, d has some value, say c, and that d increases at a uniform rate, say m units, for every unit increase in z.

Fitting a Linear Model

But you should check your hypothesis with the facts, such as they are. Should you have some values of d and z they should be plotted first of all.

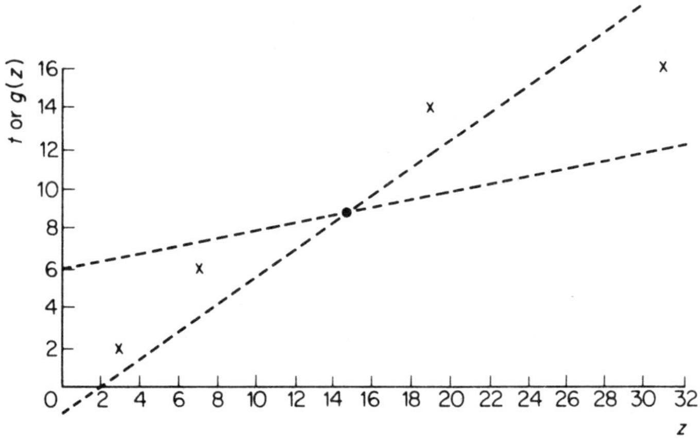

This is a plot of the values

z	3	7	19	31
t	2	6	14	16

In the diagram, the values plotted might indicate that t and z are linearly related over the range of z values from 1 to 31. Outside this range who knows what might happen. The question is, how is the line to be drawn if a linear relationship is the correct one? Many reasonable lines could be drawn by eye through these points. The $g(0)$ value could be anywhere between -1 and 6 and the slope could be from about $\frac{1}{4}$ to $\frac{1}{2}$. The problem of finding the linear relationship can be reduced to finding the best value for a and b in the function

$$g(z) = a + bz$$

where $a = g(0)$ which is obviously some value between -1 and 6,
and $\quad b =$ slope which is between $\frac{1}{4}$ and $\frac{1}{2}$.

Plot the values given in the table below

z	3	7	19	31
t	2	6	14	16

and we will use these to discuss where the line should be drawn; z will be the independent variable and will be plotted horizontally. Intuitively, the 'best' line through the points will pass through the 'centre' of the cluster, the centre being defined as the point

$$(\bar{z}, \bar{t}) \text{ or } \left(\frac{3 + 7 + 19 + 31}{4}, \frac{2 + 6 + 14 + 16}{4}\right)$$

With this point on the line fixed, the line can be rotated until it fits the data. This changes both the intercept a and the slope b; our object is to select the best pair of values a and b.

Imagine a thin strip of card pinned to your plot at $(\bar{z}, \bar{t})$ and imagine you are looking along the strip at the points. When the strip is horizontal the points seem to be well spread out, but as the strip is rotated and the position of the eye adjusted accordingly the points appear to cluster very closely about the line. In one particular position further rotation either way will increase the size of the cluster and this is the one we have been searching for.

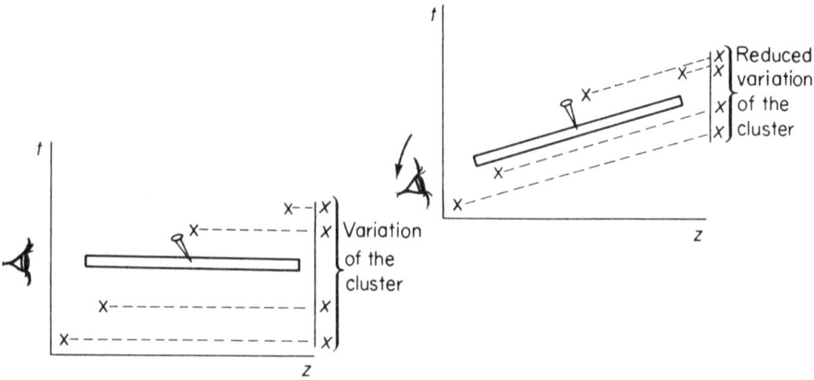

This method serves very well when there are only a few points, but when we are dealing with larger amounts of data we need a more formal procedure. Our first tentative mathematical approach might well be to add up the vertical distances of the plotted points from a trial line and rotate the line until the sum of these distances is a minimum.

This is another way of describing what happened in diagrams of the rotating line. Unfortunately, this method of choosing a line does not give us a unique solution: more than one line could be drawn which satisfies the criterion of the smallest sum.

Play with the following data. There are in fact five lines which have a total of 2 for the sum of deviations of the points from the line.

Check that their equations are

$$t = 1,\ t = z,\ t = \tfrac{1}{2},\ t = 0,\ t = 1 - z$$

by drawing them accurately.

Here is the data

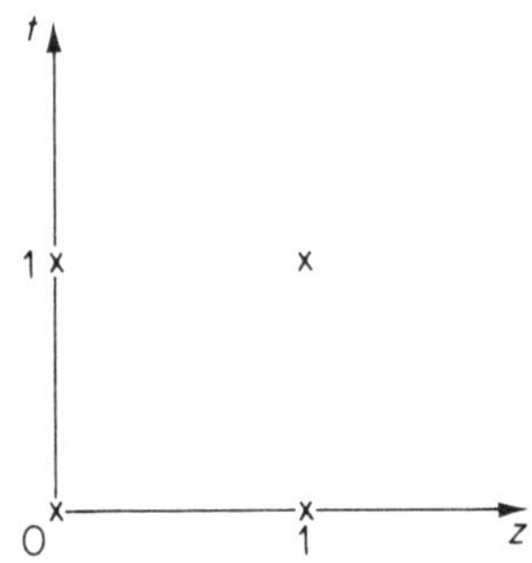

As you will have seen from the previous example, minimising the distance between the line and the points does not give us a unique solution. A unique fit is achieved by choosing a line such that the sum of the squares of the vertical differences between the values and the line is as small as we can make it. There are respectable mathematical reasons as well as uniqueness for doing this. Ask your tutor!

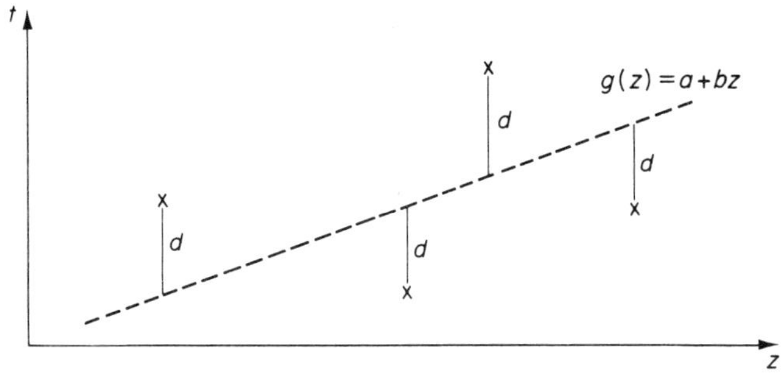

The quantity to be minimised is Σd^2, where

$$\begin{aligned}\Sigma d^2 &= \text{sum of (the actual vertical value minus the line value) squared}\\ &= (t - g(z))^2\\ &= (t - (a + bz))^2\end{aligned}$$

The problem we have to solve is to choose values for a and b such that Σd^2 is as small as possible. This time the z values and t values are known and fixed; it is the line which is being moved and a and b which change. You can try values of a and b until you select a pair which give you a value for Σd^2 which is smaller than any Σd^2 you have calculated before. But your Σd^2 will not be smaller than the one using

$$b = \frac{n\Sigma tz - (\Sigma t)(\Sigma z)}{n\Sigma z^2 - (\Sigma z)(\Sigma z)} \text{ and } a = \frac{\Sigma t}{n} - \frac{b\Sigma z}{n}$$

where n is the number of points plotted.

You should check that $\Sigma z = 60$, $\Sigma t = 38$, $\Sigma zt = 810$ and $\Sigma z^2 = 1380$ for the values in the table by drawing up a new one with headings

z, t, z times t, z squared

Substitute the values in the two formulae to get

$$b = \frac{4 \cdot 810 - 60 \cdot 38}{4 \cdot 1380 - 3600} = \frac{960}{1920} = \frac{1}{2}$$

$$a = \frac{38}{4} - \frac{1}{2} \cdot \frac{60}{4} = 2$$

The best line through the points is

$$\boxed{g(z) = 2 + \tfrac{1}{2}z}$$

If we substitute these values into the equation for Σd^2 we should get the lowest possible value for Σd^2.

The minimum for Σd^2 should be $\Sigma(t - (2 + \frac{1}{2}z))^2$.

A tabular form will simplify calculation.

① →	② →	③	④ →	⑤ →	⑥
z	$\frac{1}{2}z$	$2 + \frac{1}{2}z$	t	$t - (2 + \frac{1}{2}z)$	$(t - (2 + \frac{1}{2}z))^2$
3	1·5	3·5	2	−1·5	2·25
7	3·5	5·5	6	0·5	0·25
19	9·5	11·5	14	2·5	6·25
31	15·5	17·5	16	−1·5	2·25
					11·00

No other pair of values of a and b will give a Σd^2 value lower than 11.

Try a few easy ones! The variation about the line is minimised when the line has the equation $t = 2 + \frac{1}{2}z$.

Practice this calculation by resolving the problem of the correct line to draw through the points (0, 1), (1, 0), (0, 0) and (1, 0).

The results should not surprise you, but draw it on your sketch just the same, the practice is invaluable and instructive.

A Transformation

We have seen that it is possible to define a unique best line for a set of points which for some reason we expect to be roughly linearly related, but you will have noticed that some of the calculation is a little tedious. By moving the t axis along, the horizontal measurements of the points are changed and we can simplify the calculation. The points themselves will not move and the slope will be unchanged. All that happens is that the z value is reduced in each case by the distance moved.

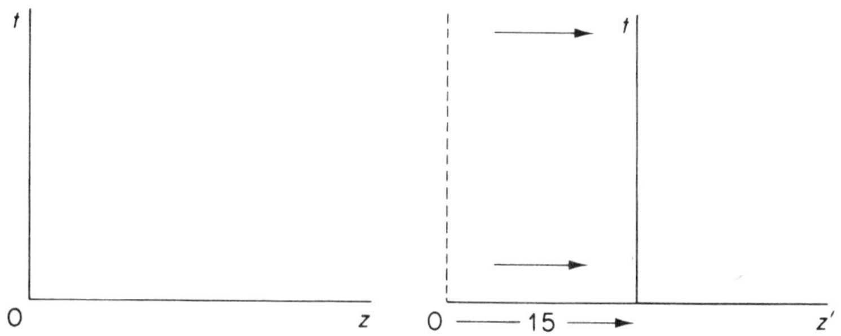

The optimum distance to move the t axis is 15, the mean value of the z so that the t axis passes through the 'centre' of the points. Calculate the equation of the 'best' straight line with the points referred to the new axes, z' and t, where z' takes the values

$$3-15,\ 7-15,\ 19-15,\ 31-15 \text{ or } -12,\ -8,\ 4,\ 16$$

and t has the values as before. Confirm that the slope is still $\frac{1}{2}$.

We could go one stage further and move the z' axis up so that it passes through the centre of the points. The t values would now become T values $2-9\frac{1}{2}$, $6-9\frac{1}{2}$, $14-9\frac{1}{2}$, $16-9\frac{1}{2}$ and the straight line would pass through the origin of the zt axes so the value of a would be 0 but the b value would remain unchanged.

Subtracting constants from the original values presented to you for analysis has no effect on the value of b, so you can subtract whatever values you like to make the calculations easy arithmetically.

There is the little matter to be considered now of remembering the formula, or at least recognising whether you have remembered it correctly. It is obviously a quotient which is symmetrical.

$$b = \frac{n \,.\, \text{sum of products} - \text{product of sums}}{n \,.\, \text{sum of products} - \text{product of sums}}$$

Both numerator and denominator have z terms in common

$$b = \frac{n \,.\, \Sigma z\ (\text{term}) - (\Sigma z)\ (\text{term})}{n \,.\, \Sigma z\ (\text{term}) - (\Sigma z)\ (\text{term})}$$

The top line terms are t's and the bottom terms are z's.

$$b = \frac{n \,.\, \Sigma zt - (\Sigma z)(\Sigma t)}{n \,.\, \Sigma zz - (\Sigma z)(\Sigma z)}$$

If you can recall the following points then you have a reasonable chance of remembering the formula correctly.

(*a*) the symmetry;
(*b*) the denominator which sounds like sigma zed-squared minus sigma-zed, squared;
(*c*) the numerator is the same as the denominator with one of the z's in each product replaced by a t.

The formula for b in most textbooks is given as

$$b = \frac{n\Sigma xy - (\Sigma x)(\Sigma y)}{n\Sigma x^2 - (\Sigma x)(\Sigma x)} \text{ or } b = \frac{\Sigma(x-\bar{x})(y-\bar{y})}{\Sigma(x-\bar{x})^2}$$

x is usually used to denote the independent variable and y the dependent variable. The choice of x and y is quite arbitrary. Practise writing the formula for variables of your OWN choice.

Thinking in graphical terms can often simplify mathematical reasoning. To illustrate this we will consider a problem which can be solved intuitively and translate it into a graphical form for solution. Once this form is established the problem can be made more complicated but the solution is automatically available on the graph.

The simple problem first of all.

A Problem and its Model

Two car-hire firms advertise different terms: Fast-car will hire at £5/day with unlimited mileage while Rollalong charges £3/day plus 2p a mile. The simple problem is to determine which firm is offering the best terms.

You can work out very quickly that for a day's trip over 100 miles, Fast-car's terms are better but a graph of cost against mileage will provide insight into the form of the problem.

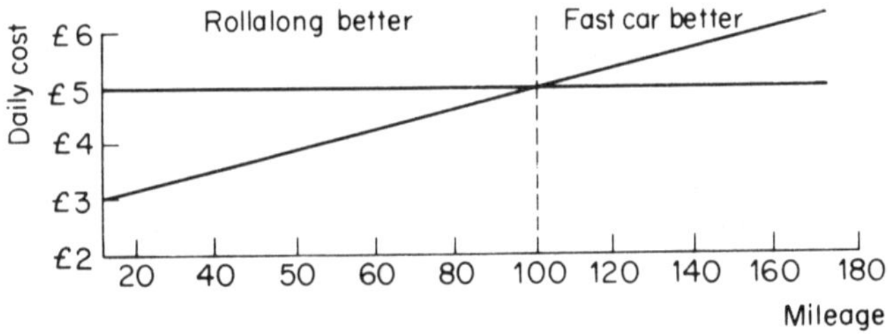

Suppose Fast-car decides, after operating their constant-rate system for some time, that their cars are suffering through over-use. To recoup their losses they impose a 3p a mile tariff for mileages over 150 miles. The arithmetic involved in solving the problem of when Fast-car vehicles should be hired now becomes more complicated, but a graph is very easily constructed to give all the information in a readily available form.

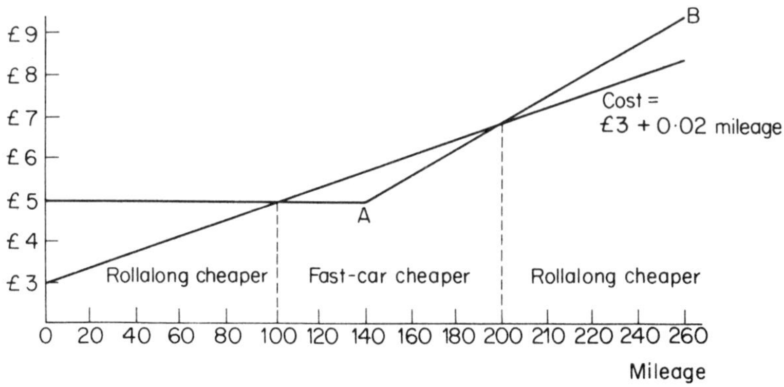

If you can now work out the equation for the line AB you will find that you have a good grasp of the basic principles of the straight-line model. Don't just sit there thinking about working it out: if all else fails, guess. Use blind guesswork to find two values *a* and *b* to substitute into the equation, cost $= a + b$ miles, which when plotted for values of from 140 to 260 miles corresponds with AB. Remember, insight is often composed of 1 per cent inspiration and 99 per cent perspiration!

You should now be able to tackle break-even analysis with complete confidence providing you can associate units of production with miles, and the overheads or set-up costs with cost of production with the daily costs.

The straight-line model for connecting two variables must be about the simplest found in business problems but using it requires a certain amount of skill in practice.

The Model Manager

A manager's attitude is all-important. What does he want from the investigation of a relationship? Confirmation of a particular prejudice? A rough idea of general trends? A detailed description for forecasting?

This attitude is often not made crystal clear when the problem is posed. Maybe the manager doesn't really know what question he should be asking or maybe he thinks that the management services section will read his mind. Perhaps he thinks that if he proposes a tightly specified problem, receives a detailed answer and then makes an inappropriate decision he will be held responsible rather than the management services department. Politics does come into mathematics in business and to ignore this fact is not to consider the problem as a whole and hence the model in the correct context. Consider the following conversations:

Conversation one:

Manager: 'Fit a straight line please.'

Reply: '$y = mx + C$!'

END of conversation!

The manager then extrapolates the y values into the future!!

Conversation two:

Manager: 'I have some data here which I have reason to think satisfies the conditions of a straight-line model.'
Management services: 'How accurate is the data?'

They establish that x is measured without error and is a controlled variable, while y is normally distributed. The management services department then produces an equation linearly relating the two variables.

Manager: 'Was there any evidence of curvilinearity?'
Management services: 'There was a quadratic element naturally but it was insignificant compared with the linear component.'
Management: 'How accurate are the values for slope and intercept?'

The management service has confidence limits for slope and intercept already calculated.

Manager: 'Can I use the model for forecasting?'
Management services: 'Certainly, if the market forces are going to be the same in the future as they were when the data was collected.'

This is a difficult qualification to meet and so they settle for the model being a rough guide and resolve to investigate relationships between other variables and the y variable.

MORAL

Solving problems is not just a question of passing messages on slips of paper or punched cards between the offices of the manager and the management services department. Each has to explain the range of the problem on the one hand and the limitations of the solution on the other.

4

Function Extended

In Chapter 3, profit was graphed against time, with time as the independent variable. Time was supposed to affect the profits in some way and their relationship was expressed as a straight line rising from left to right as the number of years increased.

The Straight-line Function

Firms keep many records besides those on profits and the Philosophers Stone in the business world is to find a variable which can be functionally related to profit: to find a single variable, x say, such as capital investment, advertising, wages and exports which could be used in an equation of the form:

Profits equal a function of x

$$P = f(x)$$

Given this relationship, the firm would only have to fix a value for the variable x and the profits P would automatically be known.

Suppose advertising was thought to be the key to profits. Increasing it would presumably bring about an increase in the profits or at least prevent them from falling. The only function we have so far considered is the linear function,

Profit = some constant + another constant times x

or

$$P = f(x) = C + mx$$

It would be a naïve businessman who thought that advertising and profits were linearly related!

Study the graph below to find out why.

Would this be a reasonable model for the relationship between advertising expenditure and profits? There are many aspects of the relationship embodied in the graph which need close investigation.

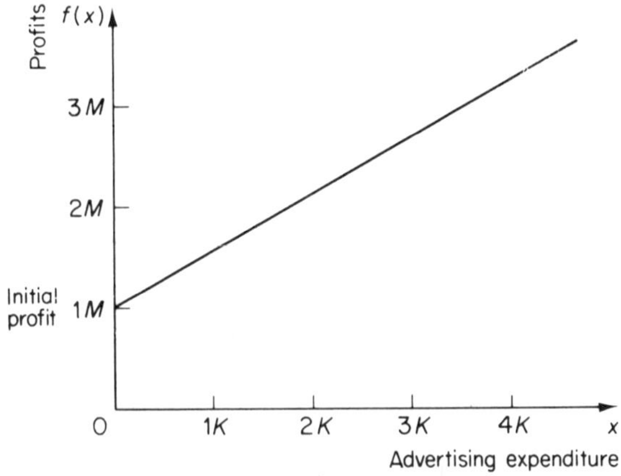

Non-linear Functions

DO we really believe, as the model indicates, that £1000 worth of advertising when none was previously employed will produce the same increase in profits as £1000 worth of extra advertising employed on a previous budget of £10*K*? Isn't it more likely that a certain amount of money must be expended on advertising BEFORE the public sees any advertisement: therefore the first few pounds spent would produce NO corresponding increase in profits? We could imagine that a few more pounds spent on the advertising budget would enable the advertising to break through into a new medium, telephone kiosks or television, with startling results on the profit.

It would be quite reasonable, though not necessarily correct, to anticipate that the relationship would follow the logic of the diagrams over the page.

The logical argument of an initial small increase in profits followed by ever-improving increases until a point is reached where no amount of increased expenditure produces increases in the profit is traced graphically in the diagrams. This is an example of the law of diminishing returns. If a firm was to pursue a policy of increasing advertising then it could reasonably expect the profits to follow the pattern developed in the diagrams.

Choosing the Function

Let us now consider the points on their own: just look at them without the connecting lines. Ignore the argument about the advertising and consider the points as a pattern on the paper and imagine that they represent data collected on profit and advertising costs for six successive years.

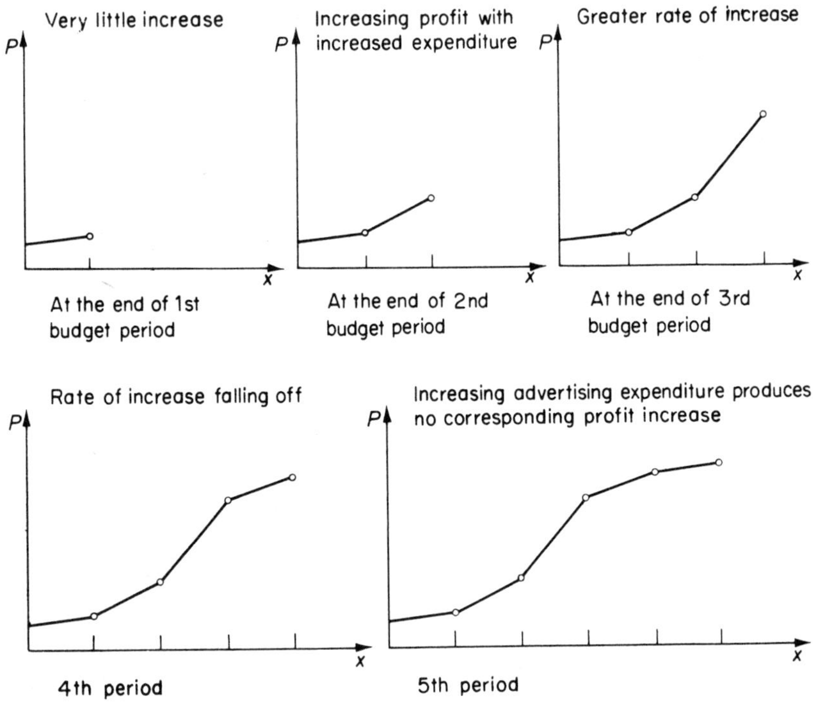

Presented with these points and no more information, an amateur mathematician could fit one of several functions to the data with startling consequences for the firm's expected profits.

Model 1. The straight line would indicate that increased expenditure would continue to mean increased profits. The firm should put all its surplus cash into advertising. Obviously absurd.

Model 2. This graph would indicate that the profits would fall to zero if much more was spent on advertising – improbable but just about possible.

Model 3. Agrees with the firm's expectations as far as they went. The firm would now have to reconsider the role of advertising in the firm in relation to other market forces now acting before putting forward

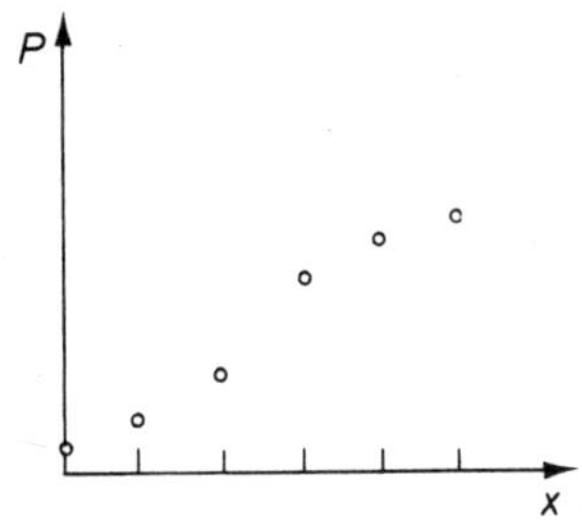

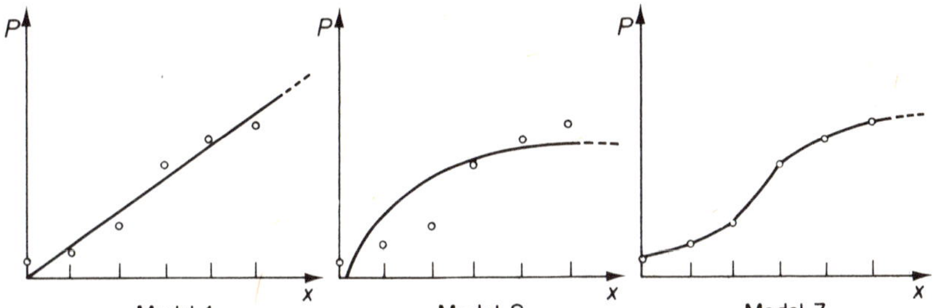

an alternative model for the future functional relationship between profit and advertising.

Careful and precise conversations need to be undertaken between management and the data analyst to determine the SHAPE of the relevant curve which describes the commonsense businessman's model of the situation. Any extrapolation from the collected data into the future depends as much on the manager's model as it does on the function provided by the mathematical analyst. Both manager and mathematician must understand enough of each other's specialised language to communicate effectively to use the data available to extract the maximum relevant information from any situation.

What ideas does the manager have to communicate to help build his model in words?

Having selected the two variables (we used profit and advertising expenditure) he decides which one he can control (the INDEPENDENT variable, advertising expenditure) and looks to see that this variable is plotted on the horizontal axis. His second variable will take different values for each value plotted on the horizontal axis and these will form a pattern which will be a straight line or a curve.

This pattern or functional relationship will generally be a simple one.

Various Functions

Here are a few examples:

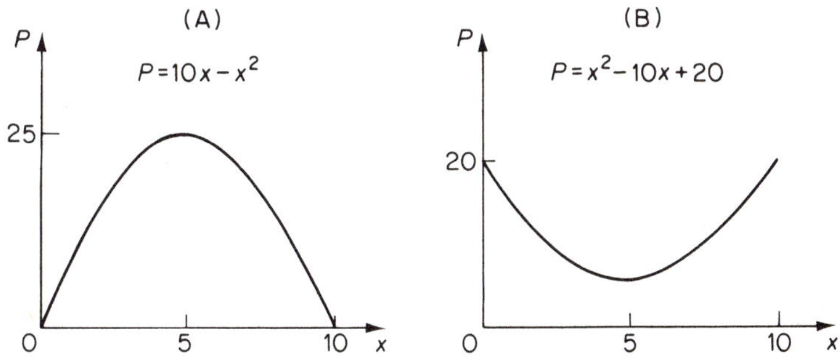

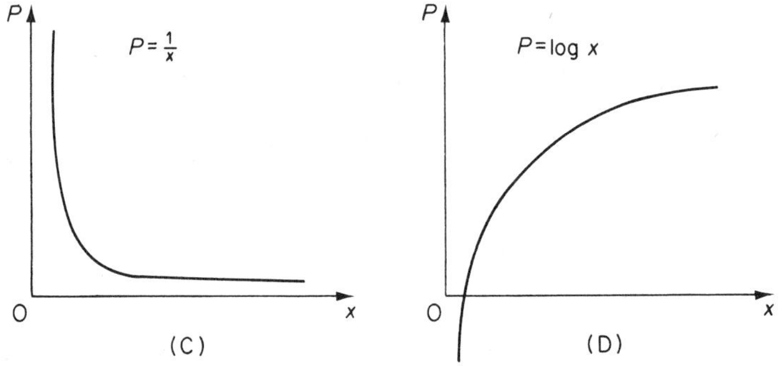

But even a simple shape such as (A) can be made even simpler by describing it as a sequence of straight lines, without losing the essential ideas contained in the shape.

This means that even very complex functions can be investigated, using only the expertise developed in Chapter 3. Let us look in detail at the function $f(x) = 10x - x^2$, x taking the values 0 to 10.

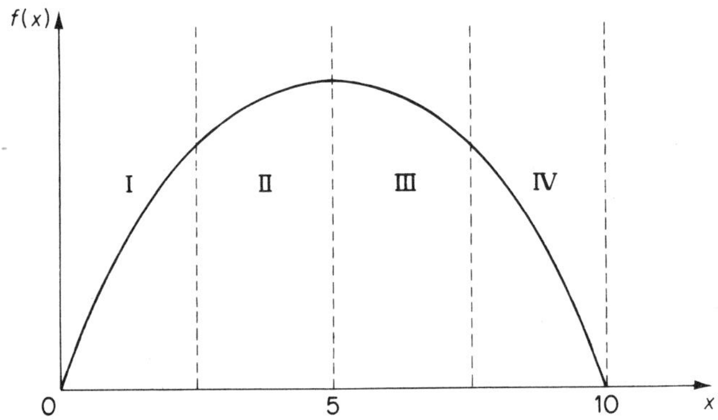

First of all we divide the range 0 to 10 into a number of regions for this function: four, I, II, III, IV, have been chosen and all have the same width.

This is not always the case. In (C) the curve is apparently made up of two straight lines connected by a curve and the intervals are very different in width.

Three regions might well be sufficient to adequately describe this curve in straight-line terms, but a more accurate representation could be obtained by dividing region II into two parts, IIa and IIb. Use this curve to practice on: plot it, and by choosing different regions see how closely straight-line approximations fit it. For $f(x) = 10x - x^2$ we

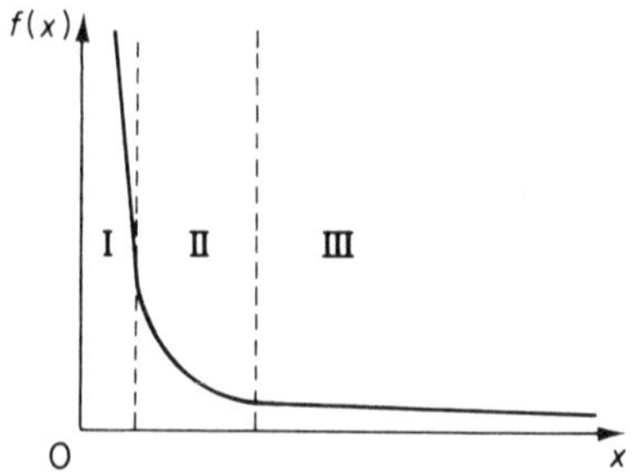

ask: what is the value of $f(x)$ at the beginning and end of the first interval?

At the beginning of interval I, $x = 0$ so $f(x)$ is

$$f(0) = 10 \cdot 0 + 0^2 = 0$$

The first point is (0, 0).

At the end of the interval, $x = 10/4$ (remember we divided the range 0 to 10 into 4 equal parts)

$$f\left(\frac{10}{4}\right) = 10 \cdot \frac{10}{4} - \left(\frac{10}{4}\right)^2 = 18 \cdot 75$$

The second point is (2·5, 18·75) so the first part of the curve is approximated by the line joining these two points. Construct a diagram with a vertical axis of about 15 cm and, using a scale of 1 unit $= \frac{1}{2}$ cm, draw on it the line connecting (0, 0) to (2·5, 18·75). Write on the line its equation.

In interval II, from 2·5 to 5, the calculation of the equation is more complicated, so we will go through the reasoning in detail. The line starts at the point calculated for the end of region I (2·5, 18·75), and ends at the point $(5, f(5))$.

From these two points plotted on the diagram the slope of this line is calculated.

Slope

$$\text{Slope} = \frac{\text{vertical change}}{\text{horizontal change}} = \frac{25 - 18 \cdot 75}{5 - 2 \cdot 5} = 2 \cdot 5$$

the equation of AB is

$$P = \text{intercept} + \text{slope times } x$$

or

$$P = \text{intercept} + 2 \cdot 5x$$

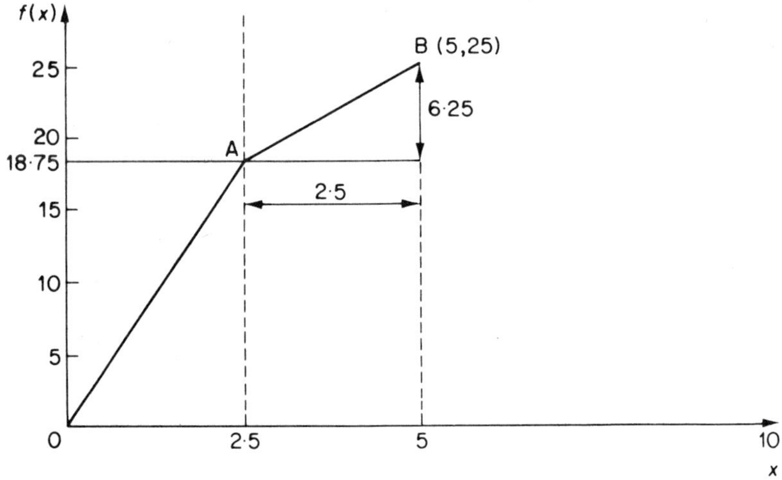

By careful drawing, the intercept is read off as 11·5 and the equation representing the straight line in region II is

$$P = 11{\cdot}5 + 2{\cdot}5x$$

This straight-line representation even has certain advantages. We can see immediately that in the region of $x = 0$ to $2{\cdot}5$, P is increasing at approximately $7{\cdot}5$/unit in x but in region II, $x = 2{\cdot}5$ to 5, the rate of increase reduces to $2{\cdot}5$, and when you calculate the equation of the straight-line approximation in the region $x = 5$ to $7{\cdot}5$ you will find that the profit now starts to decrease for increases in x.

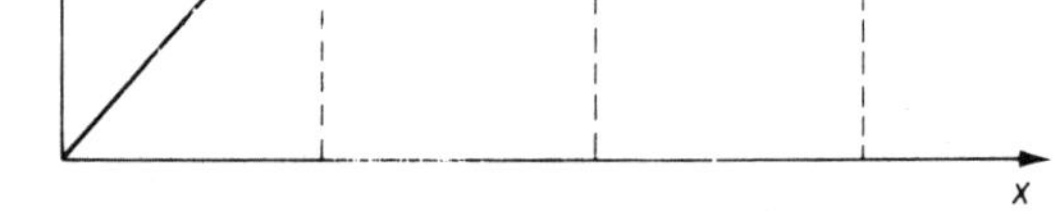

By considering the slope of the straight lines making up this functional relationship, it is easy to appreciate that there 'must' be a point where the slope is zero. This point would be pinpointed quite accurately if more straight lines were drawn. It is interesting to note that P has its largest value when the slope is 0.

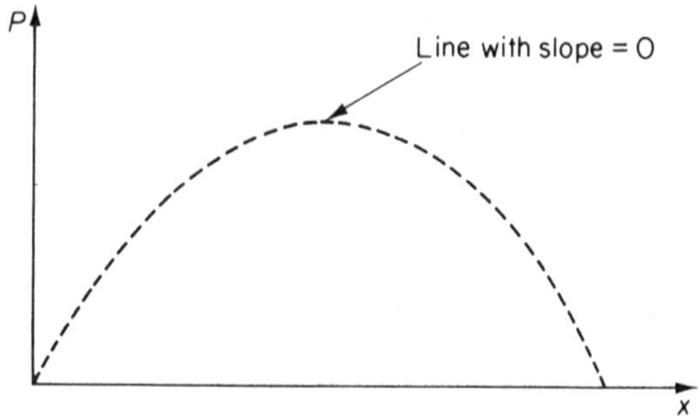

If there was a simple relationship between slope and the value x, the pattern itself would be defined. Think about it, suppose the slope were to be chosen such that it was the same for all values of x from $x =$ 0 to 10.

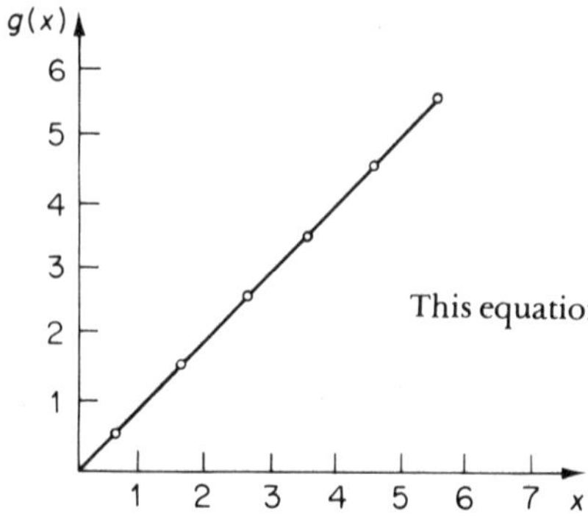

This equation, slope = 1, defines a straight-line pattern.

Even if we get adventurous and let

$$\text{slope} = x \quad \text{for all } x, \text{ from } x = 0 \text{ to } x = 10$$

we find that the pattern which emerges is quite easily appreciated.

Consider, when $x = 1$, slope $= 1$, and as x increases the slope increases, so the picture emerges through the diagrams sketched below.

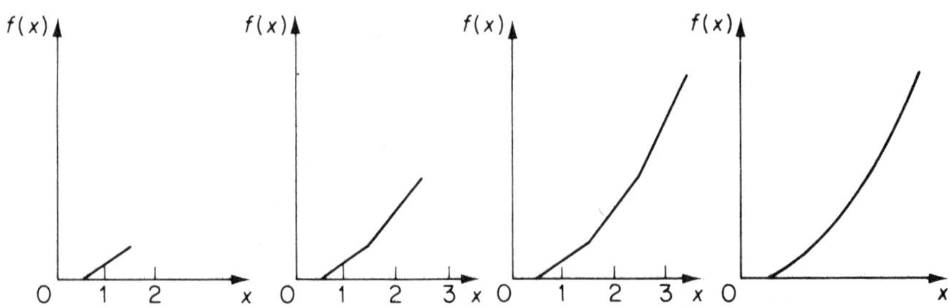

It becomes obvious that if a business situation can be described in words and can then be rephrased in terms of slopes (gradients or rates of change), then the situation can be represented graphically and the function determined. Let us now develop the idea of defining a function by describing a situation in terms of a rate of change.

Differential Equation

Imagine a company which doubles in size every ten years, forming two companies, each equal in size to the original. Each company continues to expand at the same rate and form satellite companies which continue the expansion. The more companies there are the faster the organisation grows, giving us a relationship

the more the faster

or

the rate of growth depends on the number of companies in existence at the time.

This becomes

slope = number of firms

or slope = n

Graphically, this is represented as a curve with ever-increasing slope.

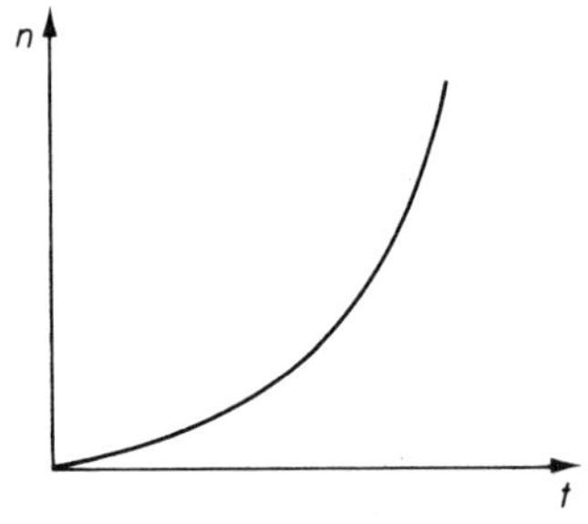

We have just written, surprisingly enough,

A DIFFERENTIAL equation

$$\frac{dn}{dt} = n$$

This equation is the distillation of all our previous assumptions and reasoning, and as such should not be read quickly and dismissed easily. It should, like all mathematics, be read slowly, taken apart and slowly reassembled, each time in a different order, so that the full implication of the relationship can be appreciated. Let us do just that.

First of all the equation has to be expressed verbally so that the speaker can be understood without resource to pencil and paper.

$$\frac{dn}{dt} = n \text{ becomes } \textit{dee en} \text{ by } \textit{dee tee} \text{ equals } \textit{en}$$

which spoken quickly, in the off-hand manner of the numerate businessman, becomes

d'en by *d'tee* equals *en*

There are two variables involved in the equation. To represent their relationship graphically we must reach a decision about the names to be given to each axis.

$$\frac{dn}{dt}$$

gives us the clue we need. It appears that something-to-do-with-n is divided by something-to-do-with-t and this ratio is the slope.

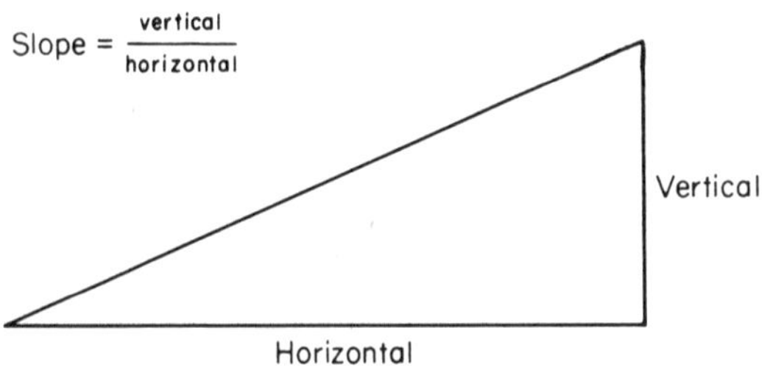

n is therefore associated with the vertical axis.

Now draw a pair of axes and label them n and t so that we can investigate how n varies with t.

When n is small, dn/dt is small, no matter what the value of t. As n increases dn/dt or the slope increases. Draw in some rough slopes for $n = 1$, $n = 2$, $n = 5$, and get the idea of the shape of the function.

As you will see from your sketch, the relationship connecting the rate of growth with n, $(dn/dt) = n$, represents an ever-increasing function. It never slows down, but rushes faster and faster into very large values of n. When n is 16 the rate of growth is 16/10/years; 10 years is the arbitrary unit of time we are using in this example.

By the end of the next 10 years there will be 32 companies and these will grow to 64 in the next 10 years.

This model, $(dn/dt) = n$ is typical of unrestricted growth situations.

The function $f(t)$ associated with it is well known and can be found

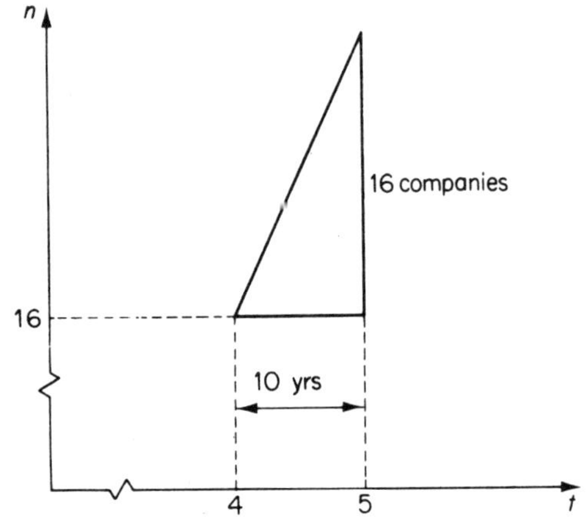

in almost any book with the title *Introductory Calculus*. One of its claims to fame is that $(dn/dt) = f(t)$; when the function is differentiated the resultant function is the same as $f(t)$. If your search for the function is unsuccessful consult your lecturer.

The function is dynamite! Just imagine what amateur demographers would predict for world population if for a particular period in history, say now, they found that the more people there were the faster the population grew, $(dP/dt) = P$! Put this way, $(dP/dt) = P$ spells DISASTER.

But, of course, this is nonsense. $(dP/dt) = P$ would only describe any sort of community for a short period of UNRESTRICTED growth. After a while pressures of food, work, standard of living would modify the original model out of all recognition. Using a model when the assumptions, in our case UNRESTRICTED growth, are no longer applicable, is called false extrapolation. Try to avoid this.

Relationships which can be expressed in terms of rates can be found in the areas of economics, accounting and engineering. Marginal cost in economics is defined as (dT/dq) say, where T is the total cost of production and q the quantity of production. In economic terms, marginal cost depends in some way on the quantity produced. In mathematical terminology we would express the rate of change of total cost as a FUNCTION of q and write

$$\frac{dT}{dq} = f(q)$$

Now $f(q)$ could be simple, $10 + 5q$ for instance, and for a few values of q between 0 and 100, $f(q)$ could be calculated and the slope sketched in to give a rough idea of the relationship between T and q. From a relationship expressed in rates of change term it is possible to quickly express the relationship in a graphical form which is much easier to interpret.

Differentiation

Given the usefulness for the businessman of finding the slope of a known function and his ability to express relationships between business variables in terms of rates, the mathematician is faced with the problem of finding techniques for calculating functions from slope relationships and, conversely, slopes from knowledge of functions.

Let us tackle first of all the problem of finding an expression for the slope of the function

$$f(x) = x^2 \quad \text{for } x \text{ from } -3 \text{ to } +3$$

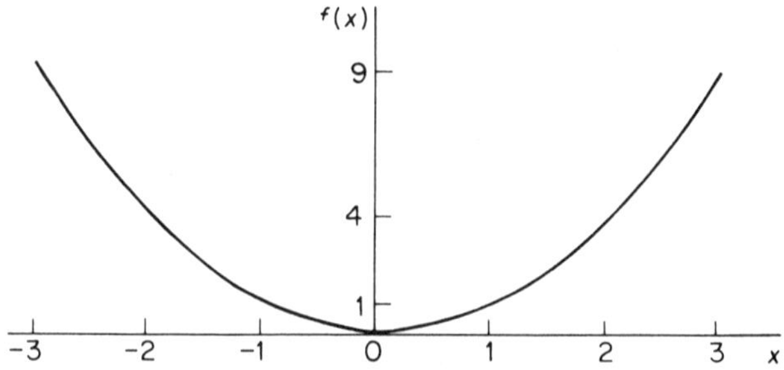

We write the slope as

$$\frac{df(x)}{dx} \quad (\textit{dee ef ex} \text{ by } \textit{dee ex}) \text{ or } f'(x) \ (\textit{ef} \text{ dash } \textit{ex})$$

This method of writing the slope provides us with a clue for discovering the method by which the slope can be approximated. The $df(x)/dx$ is a short way of writing 'the difference in the $f(x)$ value divided by the corresponding difference in the x value'.

Take as a starting point $x = -3$ and the corresponding $f(x)$ value of (-3) squared. Move to $x = -2$ and the $f(x)$ value becomes 4, a decrease of 5.

The ratio

$$\frac{df(x)}{dx} \text{ is approximately } -5 \text{ when } x = -2{\cdot}5.$$

Find the slope at $x = -2$ by calculating $f(x)$ at $x = -2{\cdot}5 - 1{\cdot}5$.

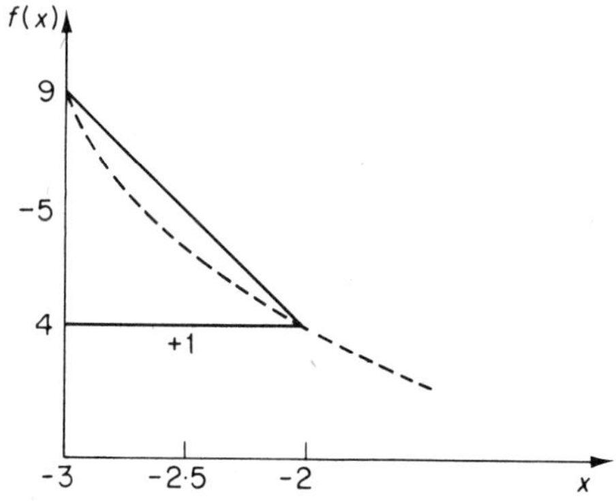

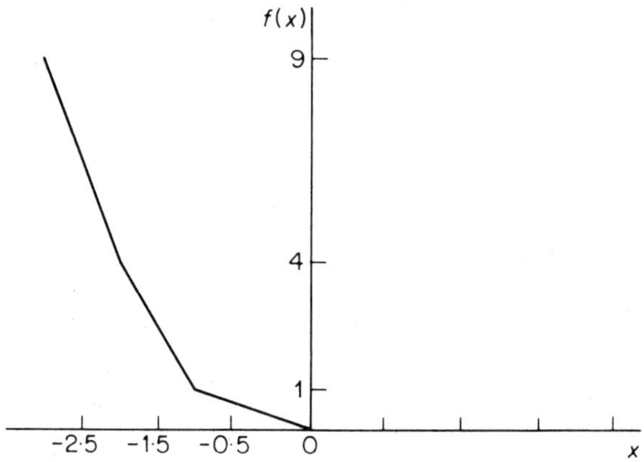

Plotting the approximate slopes against x for $x = -0{\cdot}5, -1{\cdot}5, -2{\cdot}5$ we have

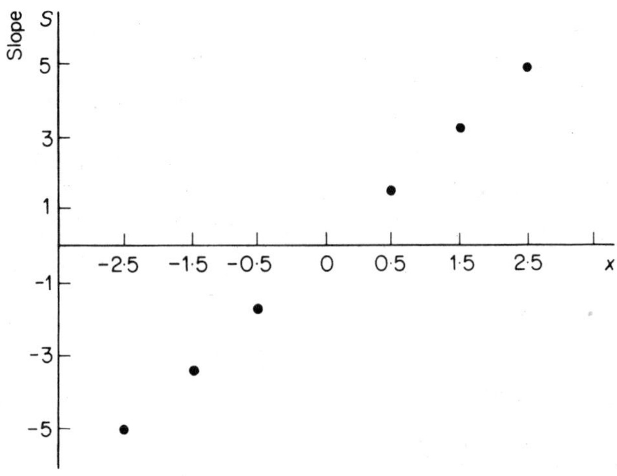

From Chapter 3 the line has the equation

$$S = 2x$$

$$\text{or} \quad \frac{df(x)}{dx} = 2x \; \frac{d(x^2)}{dx} = f'(x) = 2x$$

By consulting any book on elementary calculus you will find that the relationship between functions and their slope is given in the following table

Function $f(x)$	*Slope* $f'(x)$
x^2	$2x$
$3x^2$	$3 \cdot 2x = 6x$
x^3	$3x^{3-1} = 3x^2$
$4x^3$	$12x^2$
x^4	$4x^{4-1} = 4x^3$

For any function a businessman might reasonably use or make up, an expression can be found for the slope. Most of these are tabulated in standard books on calculus but you will need to train yourself to learn some techniques for getting from function to slope, or differentiating functions. Your lecturer is the best person to help you with this area of your mathematical development and can point out the different types of function and the methods of differentiation appropriate to each.

Inability to master these techniques of differentiation is no excuse for not solving problems involving slopes, however; a graphical method will usually provide an answer which is approximately correct.

Discussions about slopes and functions are all very well but are they really useful? Can the ideas be used to solve a real problem? In Chapter 3 we were confronted with the real problem of fitting a straight line through a series of points and, although it was solved, the reason for the solution in the form it was given was not explained. The general method for solving this problem can be understood by working through the following example.

A maintenance engineer is interested in the rate of wear of one of his machine components. This component is removed from the machine every month, weighed and replaced. The engineer finishes up with the data below

Weight (*mg*)	100	98	97	97	95
Month	0	1	2	3	4

Obviously the rate of weight loss is somewhere between 2 mg/month and 1 mg/month.

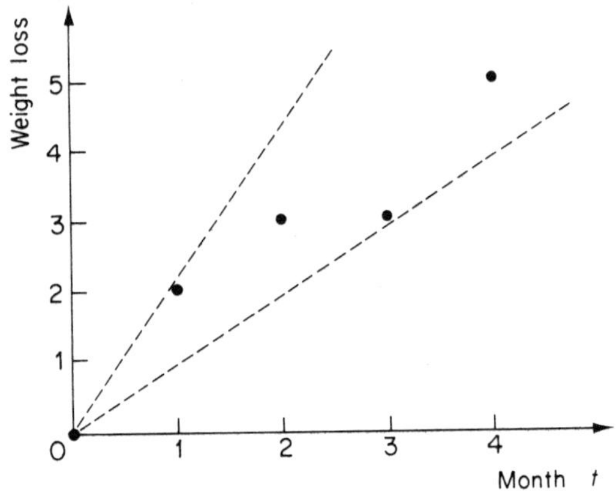

The problem is to fit a straight line through the points which also passes through (0, 0). Its equation will be of the form

$$w = \text{slope} \times t \text{ where slope is the rate of weight loss}$$

If the slope is to have a value b say, our problem is to find b such that when the line is plotted, the (differences)^2 between the weight loss predicted by the line equation and the actual weight losses is a minimum. The predicted values are b, $2b$, $3b$, $4b$, and the actual losses are 2, 3, 4, and 5.

First of all we draw any old line through the origin

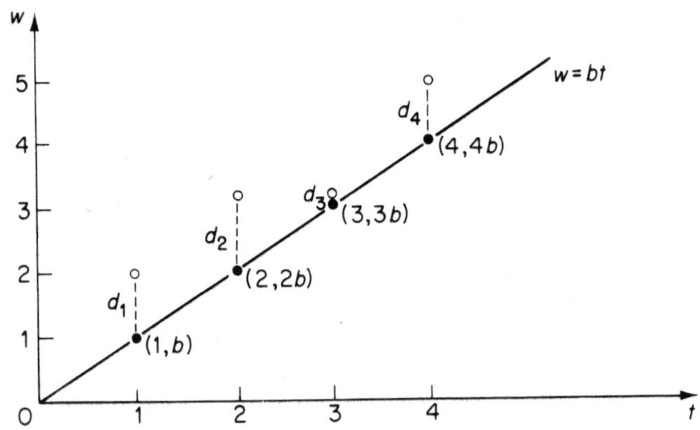

The sum of the squared distances, Σd^2, we will call S,

now $\qquad S = (2 - b)^2 + (3 - 2b)^2 + (3 - 3b)^2 + (5 - 4b)^2$

so $\qquad S = f(b)$

$\qquad S = 47 - 74b + 30b^2$

The sum of the distances squared, S, must be made as small as possible if the line is to 'fit' the points closely; b must be chosen to make S a minimum.

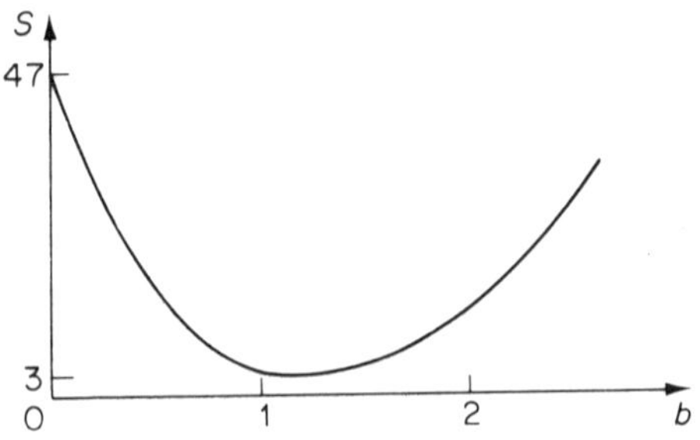

By checking the values of S when b is 0, 1 and 2 it appears that S is smallest when b is about 1·3. Somewhere between $b = 1$ and $b = 2$ the slope of S must be 0.

Consider the slope

$$\text{slope} = \frac{dS}{dt} = -74 + 60b$$

The slope is zero when $-74 + 60b = 0$, and the value of b is 74/60 or 1·266.

The engineer now has his solution derived by calculus and confirmed by eye. The rate of weight loss is 1·267 mg per month, to the best of his knowledge.

A Realistic Problem

Now we have an illustration of the graph method, we can turn to a simple but realistic problem worthy of our efforts to solve it. Of course, the problem we shall tackle was solved years ago but it has the added attraction of being quickly verified at the nearest supermarket. Quite simply, the problem is one of packaging: a manufacturer wishes to pack his paint into cylindrical tins with a circular cross section but has the problem of deciding on the shape of his can. Should it be flat like a pancake or long like a tube? It transpires that by judiciously choosing the correct shape he can save money on the amount of sheet metal used in the construction. We will investigate a can designed to hold 100 cc. The formula for the volume of a cylinder can be found in most elementary arithmetic books and often in logarithm tables.

Usually the formula is written as

$$V = \pi r^2 h$$

or: volume = 3·14 × radius × radius × height.

Our volume is fixed so

$$100 = 3{\cdot}14 r^2 h$$

and by dividing both sides of the equation by $3{\cdot}14r^2$ we get

$$h = \frac{100}{3{\cdot}14 r^2}$$

In our previous terminology, $h = f(r)$; for every value of r there is a value of h.

Now we have used the volume to connect h and r, we consider the area of metal involved.

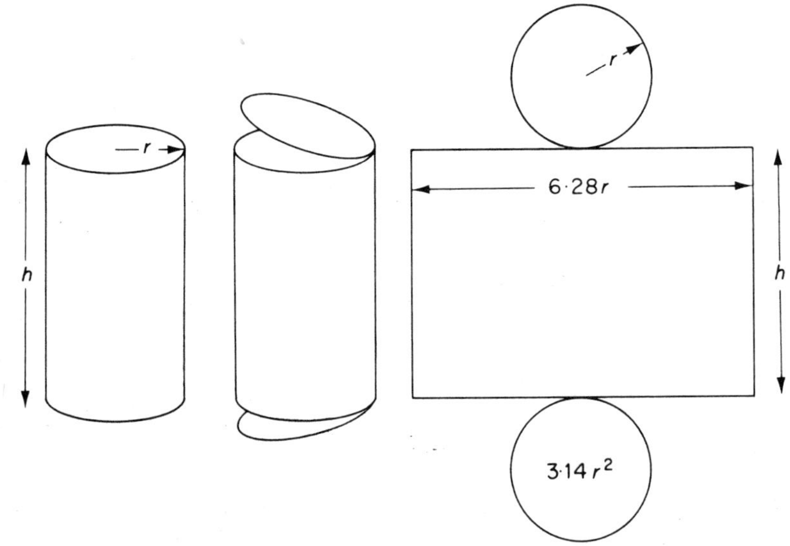

The surface area of the cylindrical can is made up of the cylinder, with area $6{\cdot}28rh$, and the two circular ends each with area $3{\cdot}14r^2$.

The total area = $6{\cdot}28rh + 3{\cdot}14r^2 + 3{\cdot}14r^2$.

Surface area can now be written

$$S = 6{\cdot}28\,(rh + r^2)$$

Unfortunately, S cannot be plotted against r or h because as one changes so does the other, to keep the volume constant at 100.

$$h = \frac{100}{3 \cdot 14r^2} \qquad \text{Remember?}$$

With the information from the volume formula, S can be expressed in terms of r alone.

$$S = \frac{6 \cdot 28 \,.\, r \,.\, 100}{3 \cdot 14 \,.\, r^2} + 6 \cdot 28r^2 = f(r)$$

Simplifying this expression we have

$$S = \frac{3 \cdot 14r}{3 \cdot 14r} \cdot \frac{2 \,.\, 100}{r} + 6 \cdot 28r^2$$

$$S = \frac{200}{r} + 6 \cdot 28r^2$$

It is difficult to visualise the function but by plotting S for r values between 1 and 8 it is possible to discern that S is a 'U'-shaped function and U-shaped functions have a lowest point. Our problem is to find a value for r which, when substituted into the equation for S, gives the smallest surface area. At this value for r, the slope ds/dr will be zero.

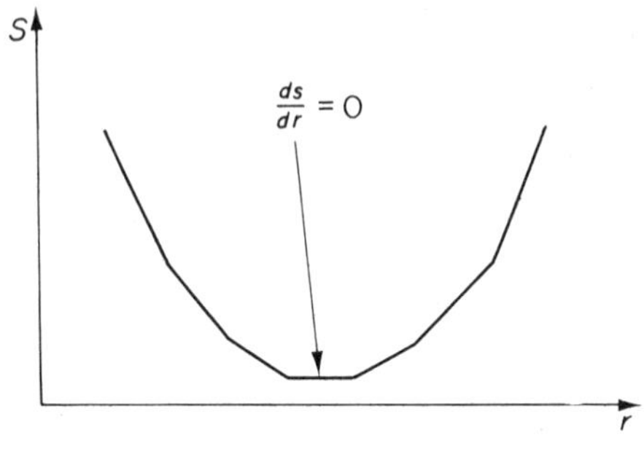

ds/dr is easily found.

Noting that $1/r$ can be written r^{-1} we have $(ds/dr) = -200r^{-2} + 12 \cdot 56r$.

When the slope is zero, S will be a minimum.

Putting $(ds/dr) = 0$ we have, $0 = -200r^2 + 12 \cdot 56r$, which can be reorganised to give $12 \cdot 56r^3 = 200$, so that

$$r^3 = \frac{200}{12 \cdot 56}$$

r could be calculated from this equation but we really need to know the relationship between r and h; so we return to the volume equation

$$100 = 3{\cdot}14r^2h$$

Now we know that

$$r^3 = \frac{200}{12{\cdot}56}$$

so that

$$r^2 = \frac{r^3}{r} = \frac{200}{12{\cdot}56\,.\,r}$$

$$\text{and} \quad 100 = 3{\cdot}14\left(\frac{200}{12{\cdot}56\,.\,r}\right)h$$

therefore $h = 2r$.

For optimum use of the metal, the diameter of the can should be the same as the height! Note: A very satisfying result. Had you not sketched the function, solving $(ds/dr) = 0$ might have given you the position of a maximum rather than a minimum. Mathematics teachers usually insist that this possibility is always investigated and quite rightly but, as has been mentioned before, a sketch will usually clarify the position.

The slope has many uses: it can give insight into other aspects of problem-solving using mathematics.

Everyone knows how to solve simple problems of the following type: the revenue from the sale of 500 double beds is £25,000, what is the cost of one bed?

No one would have any difficulty deducing that the cost is £50/bed, but the method of reaching that answer is what concerns us now. Maybe other problems, possibly more difficult but of the same type, could be solved by this method.

The revenue in the problem can be written two ways:

£25,000
and 500 × cost of a bed.

If we replace cost of a bed by c and write

$$25{,}000 = 500c$$

the value for c almost jumps out of the equation at you, but restraining it for further investigation the equation can be rewritten

$$25{,}000 - 500c = 0$$

and the problem is now to find a value for c which makes the left-hand side of the equation zero.

Experimenting with a few values of c we have

c	$25{,}000 - 500c$
1	$25{,}000 - 500 = 24{,}500$
10	$25{,}000 - 5{,}000 = 20{,}000$
4	$25{,}000 - 2{,}000 = 23{,}000$

Apparently, $25{,}000 - 500c$ is a function of c, let us say $g(c)$ and we must now solve $g(c) = 0$.

Graphically, this is easy: $g(c) = 0$ where the line representing the function crosses the c axis.

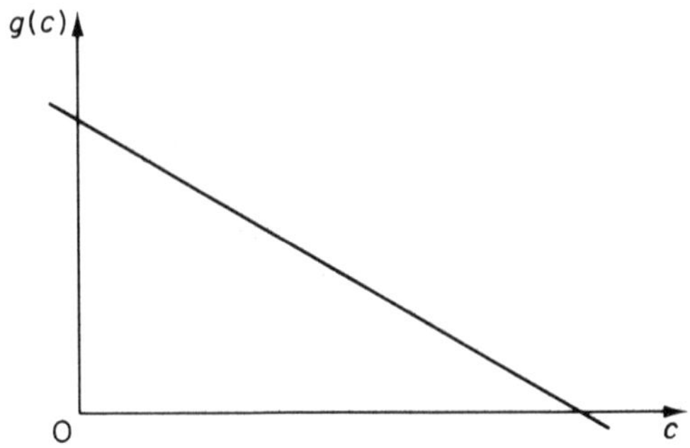

$25{,}000 - 500c$ is an easy function to plot and the value of c is quickly read off as 5.
BUT
suppose, just suppose, our problem reduced to a more complicated equation of the form

$$f(x) = x^2 - 6x + 5 = 0$$

$f(x)$ is not a straight line but it could still be plotted to find where $f(x)$ crosses the x axis, where $f(x) = 0$.

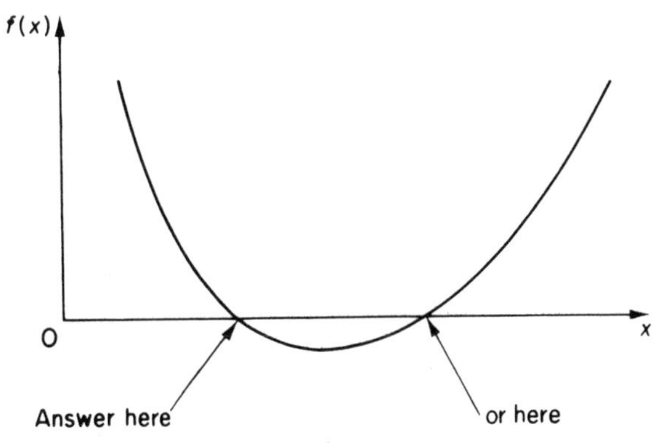

The problem, you will have noticed, is just to find a set of reasonable values of x to plot $f(x)$.

A bad selection could lead to a futile plot, one which does not cross the axis.

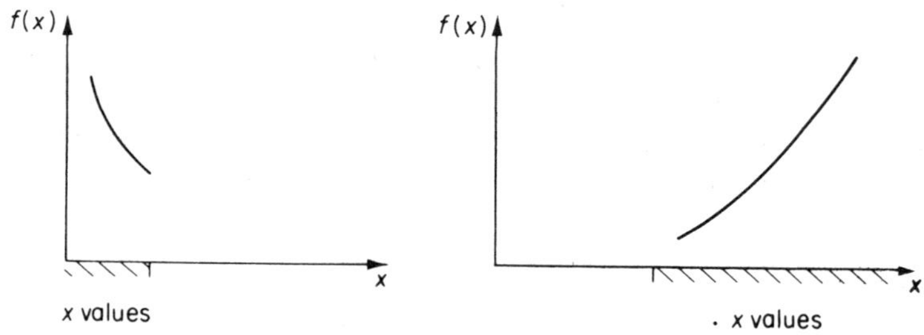

The chance of choosing the wrong set of values of x to start with is high so we could do with a method of starting with the wrong x values and quickly jumping to the area of interest.

For example, suppose we guess wildly that $x = 10$

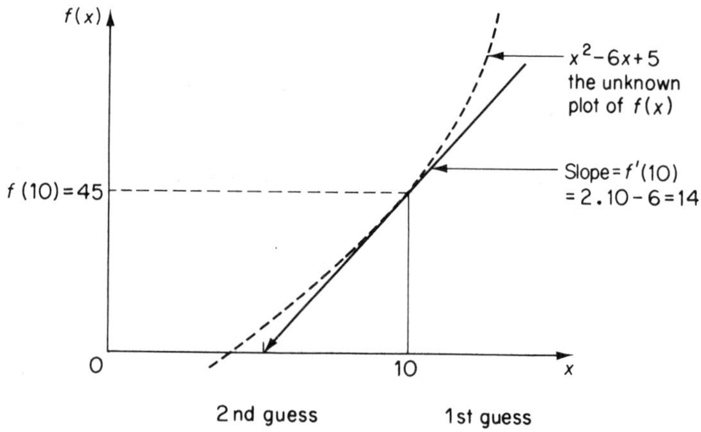

and draw a line with the same slope as the function has at $x = 10$, then this line would meet the x axis nearer to the place where $f(x) = 0$.

The line meets the x axis $45/14$ to the left of 10.

Since we are solving this problem in an experimental frame of mind, $45/14$ is roughly 3, so the line meets the axis at about $x = 7$.

The slope of the curve at $x = 7$ is $2 \times 7 - 6 = 8$ and

$$\text{slope} = \frac{\text{vertical distance}}{\text{horizontal distance}} = \frac{12}{h}$$

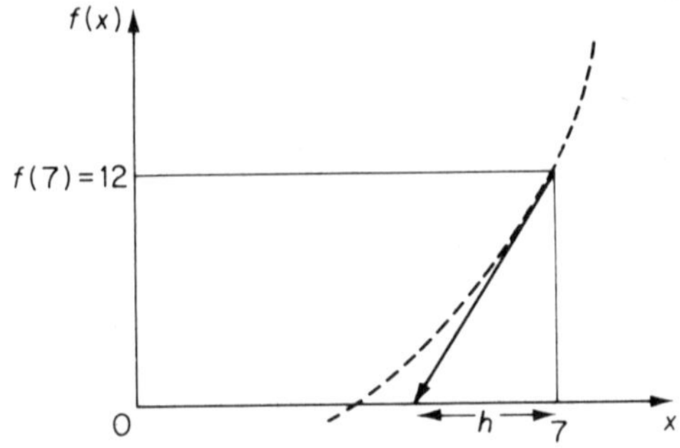

So $h = 1\frac{1}{2}$ and we now have a closer estimate of x at $f(0)$.
The second estimate of x at $f(0) = 5\frac{1}{2}$.
We try $x = 5$ and see if that guess can be improved. At $x = 5, f(x) = 0$!

From a wild guess of $x = 10$ we have arrived at the answer in three steps; we have 'converged' systematically on one of the values of x which satisfy the equation.

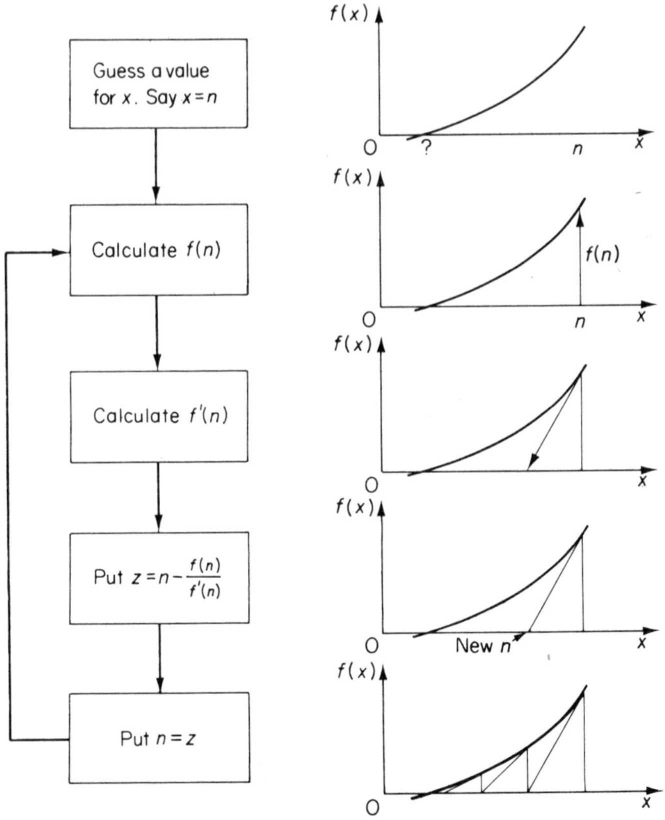

Algorithm for solving $f(x) = 0$

There is another answer. Try to find it by starting with a wild guess of 0 and do not be put off by negative slopes.

Continue your investigation by starting with a wilder guess of $x =$ 100 and see how long it takes to converge to the correct answer. Is there a limit to the wildness of a first guess?

Experiment! The answer will only come through calculation. Unfortunately, mathematical thinking is not caught like measles but grows slowly like a callous, the result of arithmetic spadework. To help with this spadework necessary to grasp the concepts of iteration and convergence we have considered, the flow diagram on the previous page describes the method or algorithm we use.

Each time the loop of instructions is completed, the new n gets closer to the value of x at $f(0)$, but we are now faced with the obvious question: when do we stop? When we are tired? When we are close enough? If the answer to this problem does not come easily get a maths book which has in its index a topic called 'Newton's method' or 'Newton Raphson method' and find out. The book might be full of incomprehensible symbols but you can still use it to find the answer to your question. If you are not that adventurous, ask your lecturer!

Mastering this method of solving $f(x) = 0$ is well worthwhile for two reasons: firstly, many, many problems reduce to the solution of $f(x) = 0$; and secondly the method works for almost any function, provided of course it can be differentiated to find its slope. Do not worry about the time it seems to take because it is well spent, and once understood the method can be easily made into a computer program whereupon the time of solution is trivial.

5
Modelling Systems

Equations as Models

What does an equation mean? In the earlier pages of this book, if you saw an equation like

$$y = 2x + 3,$$

or

$$P = 1000x + 500$$

then the emphasis was on FINDING y or P from particular values of x. However, an equation has much more to it than that: they need not be SOLVED; they can just BE. An equation represents the relationship between two (or more) variables. That relationship could be expressed verbally, e.g. 'When x gets larger, P gets larger'.

One aim of this book is that you should be able to write down equations from verbal statements but, even more important, you should be able to translate the equations into words and appreciate what the equation represents. 'When x gets larger, P gets larger.' You can see this happening in the following diagram.

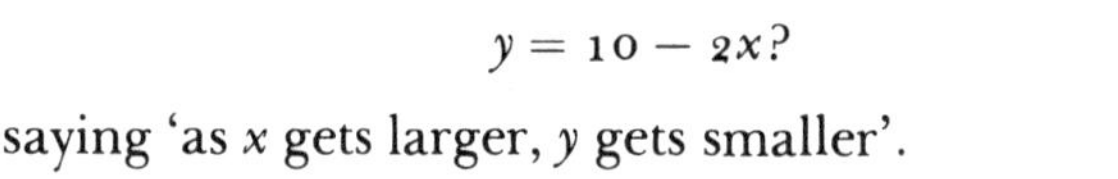

Do you really understand the meaning of

$$y = 10 - 2x?$$

This is saying 'as x gets larger, y gets smaller'.

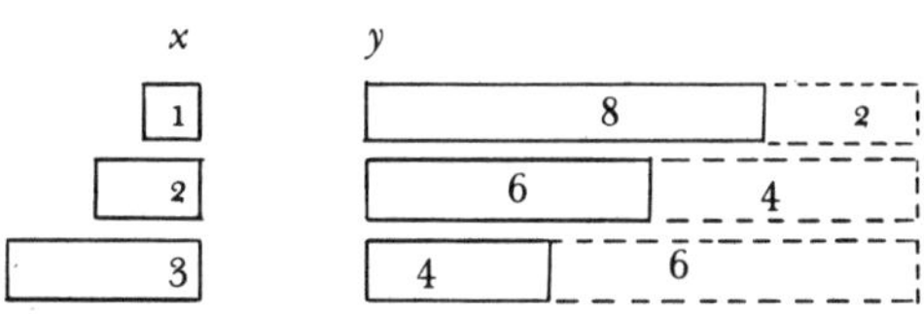

You can see y shrinking. This is obvious if you rewrite the equation

$$y + 2x = 10$$

Since y and $2x$ always add up to 10, then if one gets larger the other must get smaller.

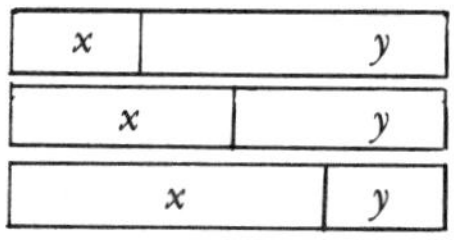

A set of equations can be used to show the relationship between several variables. If you study economics you may meet the following in relation to the theory of National Income.

$$Y \equiv C + S$$
$$O \equiv C + I$$
$$I \equiv K + G$$

The first equation says national income, Y, is equal to what is spent on consumables, C, plus what is saved, S. The second equation says that total output, O, is made up of goods that are sold, C, and goods that are produced for investment, I. Lastly, goods that are produced for investment, I, are made up of capital equipment, K, and stocks of unsold consumables, G. Each equation is a DEFINITION of a variable in terms of other variables. That is why it is written with an equivalence sign $\equiv$. Contrast with an EQUATION which is an attempt to describe a relationship between variables.

A similar set of equations may be drawn up to represent the number of guests staying in a hotel. Clearly the guests staying at a hotel on a particular night are from two sources: those who arrived previously and have not yet departed, and new arrivals.

The remaining guests, R, are made up of those who were staying yesterday less those who departed.

So, the number of guests staying tonight is

$$G = Y - D + A$$

An equation of a similar style would do for the number of guests staying tomorrow (or any other) night, but each letter would apply to a different day. This is an opportunity to use a special notation to incorporate an idea of time.

$$G_{\text{Saturday}} = G_{\text{Friday}} - D_{\text{Saturday}} + A_{\text{Saturday}}$$
$$G_{\text{Sunday}} = G_{\text{Saturday}} - D_{\text{Sunday}} + A_{\text{Sunday}}$$

Instead of actually writing the day of the week, we can use subscripts: T refers to today, $T-1$ to yesterday, $T-2$ to the day before that, etc. So

$$G_T = G_{T-1} - D_T + A_T \qquad \text{(equation 1)}$$

This is an example of the use of 'difference' notation, so called because the subscripts show the difference between the days being dealt with. This is a general equation which describes what happens every day, by the insertion of the appropriate value of T. For instance,

$$G_3 = G_2 - D_2 + A_2$$

To interpret this you need a key to show which day is which.

Key to 'T' Notation

Day	*Sun*	*Mon*	*Tues*	*Wed*	*Thurs*	*Fri*	*Sat*
T	1	2	3	4	5	6	7

The advantage of using the difference notation is that it identifies the similarity in the structure of the equation for each day by using the same equation, whilst retaining the difference between each day by using a different value of T. Without this notation a whole set of equations would have to be used for each day being analysed.

Equation 1 'looks backwards' to see where tonight's (T-night's) guests have come from. An alternative approach is to describe TOMORROW NIGHT'S ($T+1$ night's) guests. The equation

$$G_{T+1} = G_T - D_{T+1} + A_{T+1}$$

is an alternative to equation 1, and describes the typical form of another set of equations which also show what is happening.

The subscript notation can be taken further. Suppose ${}_1A_{\text{Mon}}$ is the number of guests who arrived on Monday and stayed 1 night, ${}_2A_{\text{Mon}}$ represents those who stayed 2 nights, ${}_3A_{\text{Mon}}$ for those who stayed 3 nights etc. The first subscript refers to the length of the visit, the second subscript to the arrival date, so that ${}_2A_{\text{Tues}}$ is the number of guests who arrived on Tuesday and stayed 2 nights. The following table uses this notation to categorise the guests.

Table of *A*'s

		Length of visit (days)					
	Day	1	2	3	4	5	etc....
	Mon	${}_1A\text{Mon}$	${}_2A\text{Mon}$	${}_3A\text{Mon}$	${}_4A\text{Mon}$	${}_5A\text{Mon}$	...
	Tues	${}_1A\text{Tues}$	${}_2A\text{Tues}$	${}_3A\text{Tues}$	—	—	...
	Wed	${}_1A\text{Wed}$	${}_2A\text{Wed}$	${}_3A\text{Wed}$	—	—	...
Day of Arrival	Thurs	${}_1A\text{Thur}$	—	—	—	—	...
	Fri	${}_1A\text{Fri}$	—	—	—	—	...
	Sat	—	—	—	—	—	...
	Sun	—	—	—	—	—	...
	Mon	—	—	—	—	—	...
	etc.						
	.						
	.						
	.						

Take out a piece of paper and make out a complete table for yourself and use it for reference when reading the next few pages. Notice how each row refers to all the guests who arrive on a particular day; for example, the first row refers to Monday's arrivals, and groups them according to how long they stay. The TOTAL number of guests who arrive on Monday is made up of the ONE nighters, the TWO nighters, the THREE nighters, etc.

TOTAL = the ONE nighters + the TWO nighters + etc.

We already have an expression for Monday's arrivals, A_{Mon}, and so now

$$A_{\text{Mon}} = {}_1A_{\text{Mon}} + {}_2A_{\text{Mon}} + {}_3A_{\text{Mon}} + \text{etc.}$$

This equation is made up of the first row of the table and shows the breakdown of Monday's guests. In Chapter 2 this breakdown was investigated. The actual figures were as follows:

$$86 = 31 + 21 + 9 + 19 + 3 + 0 + 0 + 0 + 0 + 2$$

The behaviour of guests who arrive on other days is quite different but similar equations can be used to represent it. For example, the breakdown of Saturday's guests is

$$A_{\text{Sat}} = {}_1A_{\text{Sat}} + {}_2A_{\text{Sat}} + {}_3A_{\text{Sat}} + {}_4A_{\text{Sat}} \text{ etc.}$$

which uses another row of the table.

A general equation can be written to represent the composition of guests who arrive on a general day, 'T'-day,

$$A_T = {}_1A_T + {}_2A_T + {}_3A_T + \ldots \qquad \text{(equation 2)}$$

This equation has the FORM of all the others but, of course, the VALUES of the A's for different days are quite different.

Another general equation is the following:

$$D_T = {}_1A_{T-1} + {}_2A_{T-2} + {}_3A_{T-3} + \ldots \qquad \text{(equation 3)}$$

Can you interpret it using your table? D_T is the total number of guests who DEPARTED on T-day. When interpreting equations it is often helpful to consider a particular case; for example, if T-day is Sunday then the equation says

$$D_{\text{Sun}} = {}_1A_{\text{Sat}} + {}_2A_{\text{Fri}} + {}_3A_{\text{Thurs}} + {}_4A_{\text{Wed}} + \ldots$$

This shows the breakdown of guests who departed on Sunday. Some of them will have arrived on Saturday and stayed only 1 night, some will have arrived on Friday and stayed 2 nights, and so on, into the past. The particular groups of guests, the A_{Sat}, A_{Fri} etc., form a DIAGONAL of the table. We don't know how far into the past to take the diagonal because we cannot specify how LONG a guest, departing on Sunday, could have stayed. It could be 2 weeks, it could be a year.

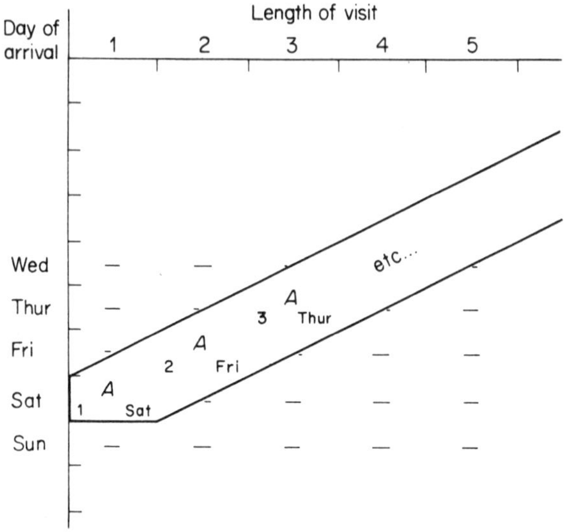

Write down an equation to show the breakdown of Thursday's departures and construct the corresponding diagonal in your table. The interpretation of the general equation showing the breakdown of guests, who departed on T-day, is now obvious. Some will have arrived the day before and stayed only 1 night, some will have arrived two days ago and stayed 2 nights and so on.

The general equation 3 is important for it is relevant to the problem of determining the number of sheets required by the hotel. Since sheets are changed when a guest leaves (we hope!), then

$$2 \times D_T$$

is the number of clean sheets required for the D_T vacated beds. Reminding you once more of the hotel's policy of changing sheets every 4 days of a guest's visit, then sheet changes are also required on T-day for the guests who arrived 4 days ago, 8 days ago, 12 days ago, etc. If we assume all beds in unoccupied rooms have clean sheets, we don't need to consider the number of guests arriving on day T.

$$\text{So the total sheets changed today} = 2 \times \begin{pmatrix} \text{today's} \\ \text{departures} \\ + \\ \text{guests who} \\ \text{arrived 4 days} \\ \text{ago and} \\ \text{haven't left} \\ + \\ \text{guests who} \\ \text{arrived 8 days} \\ \text{ago and} \\ \text{haven't left} \\ + \\ \text{guests who} \\ \text{arrived 12 days} \\ \text{ago and} \\ \text{haven't left.} \end{pmatrix}$$

But guests who arrived 4 days ago and haven't left must be staying MORE than 4 days ('4+' days). Using the difference notation again we get an equation called MODEL 1.

Guest Activity Models

MODEL 1

$$\text{Sheets changed}_T = 2(D_T + {}_{4+}A_{T-4} + {}_{8+}A_{T-8} + {}_{12+}A_{T-12} + {}_{16+}A_{T-16} + \ldots) \qquad \text{(equation 4)}$$

This notation can get very complicated, but the ideas behind the equation are very simple. The equation represents a set of equations referring to EVERY day. Once more the meaning becomes clearer when the equation relating to a PARTICULAR day is considered. For example,

$$\text{Sheets changed}_{\text{Sun}} = 2(D_{\text{Sun}} + {}_{4+}A_{\text{Wed}} + {}_{8+}A_{\text{Sat}} + {}_{12+}A_{\text{Tues}} + \ldots)$$

The table of A's shows the particular groups of guests who are due for sheet changes on Sunday.

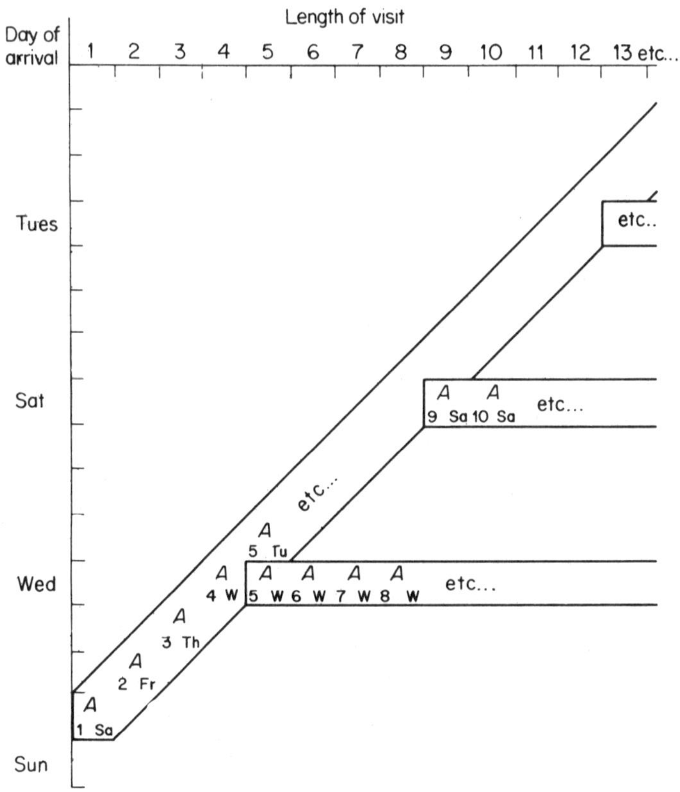

Detailed breakdown

A less detailed diagram on the next page identifies the terms in the equation.

You can construct a similar diagram to represent the terms in the general equation 4.

If the hotel works exactly to plan, then the equation describes exactly what will happen but, bearing in mind the work involved each day in establishing how long each guest has been staying and whether he is 'due' for a sheet change, practice is probably not in tune with policy. Therefore the equation 4 (or rather the SET of equations) is an attempt to describe what happens, so can be used as a model of sheet-change behaviour (MODEL 1).

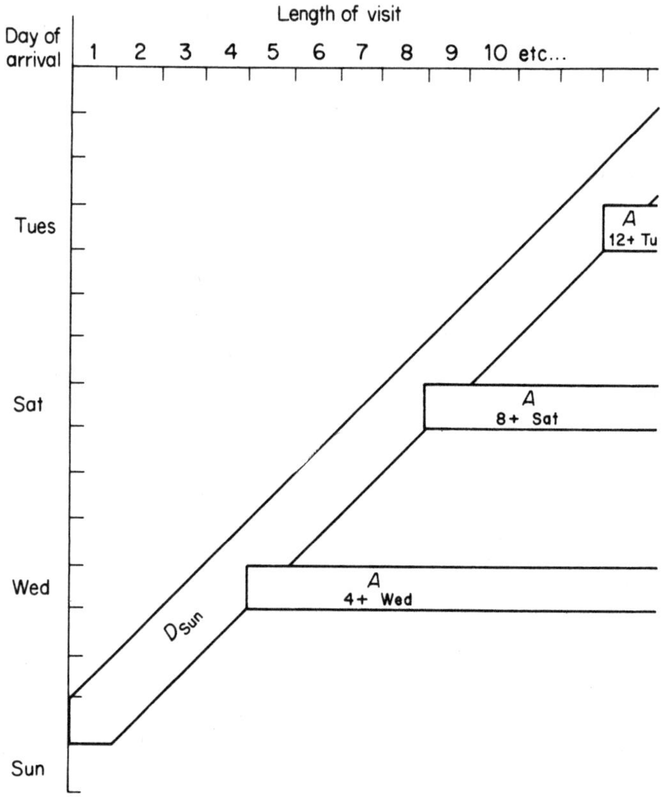

You might think the model is too elaborate since it considers every possible sheet change. Those for guests staying more than 16 days can be neglected with little loss of accuracy, because so few stay this long. So the number of sheet changes on day T, (SC_T), can be expressed as,

$$SC_T = 2(D_T + {}_{4+}A_{T-4} + {}_{8+}A_{T-8} + {}_{12+}A_{T-12})$$

Or, again, if the number of guests staying over 12 days is small, you might settle for,

MODEL 2

$$SC_T = 2(D_T + {}_{4+}A_{T-4} + {}_{8+}A_{T-8})$$

as a reasonable model.

Other models can be developed which may be closer to actual behaviour rather than policy. The domestic staff may consider the problem of determining which guests are due for sheet changes as too awkward and interpret policy instead by changing all beds (for occupied rooms) every 4 days. This is not the same thing by any means.

For instance, it might mean that a guest intending to stay for 3 nights has his sheets changed after 1 night because the 'sheet-changing' day occurred then.

In order to model this new sheet-changing behaviour it is useful to consider the three types of guests with regard to T-day:

(1) The guests who DEPART on T-day, D_T.
(2) The guests who REMAIN on T-day, $G_{T-1} - D_T$.
(3) The guests who ARRIVE on T-day, A_T.

The following diagrams illustrate the activity on T-day.

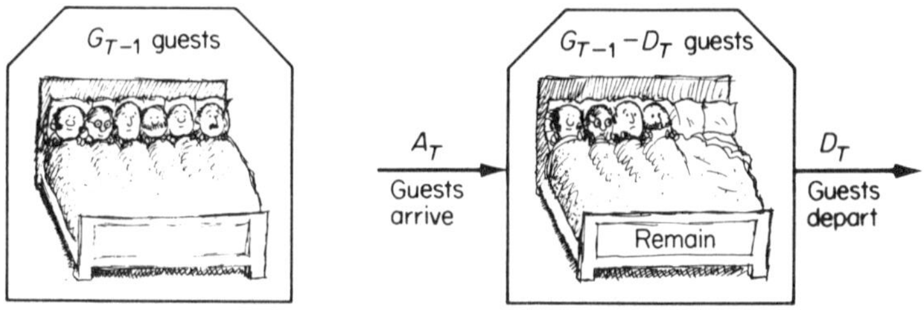

Very early T-day morning   During T-day

The equation relating G_T and G_{T-1}, the numbers of guests staying on successive nights, is

$$G_T = G_{T-1} - D_T + A_T$$

Yesterday, ($T - 1$ day), there were G_{T-1} guests in the hotel, so on T-day morning there are G_{T-1} dirty pairs of sheets to be changed.

If T is a 'sheet-changing' day (every fourth day) then all these will be changed. In the intervening days the beds of the guests who depart (D_T of them) must be changed to leave the rooms in a fit state for new arrivals. The model is therefore as follows:

Hotel Activity Models

MODEL 3

$$SC_T = 2G_{T-1} \text{ if } T \text{ is a 'sheet-changing' day}$$
$$SC_T = 2D_T \text{ if } T \text{ is a routine day}$$

The first equation represents the major sheet changes occurring every fourth day and the second equation represents the routine work of the other three days of the four-day cycle.

Use your table of A's to identify the groups of guests who will have their sheets changed on a 'sheet-changing' day. You will find it helpful to consider the composition of the G_T guests who stay on T-night, which can be seen from the following equation:

$$G_T = \left(\begin{array}{l} \text{all the arrivals} \\ \text{on } T \text{ day} \\ + \\ \text{all those who} \\ \text{arrived on } T\text{-}1 \\ \text{day who haven't left} \\ + \\ \text{all those who} \\ \text{arrived on } T\text{-}2 \\ \text{day who haven't left} \\ + \\ \text{etc. . . .} \end{array} \right)$$

$$G_T = A_T + {}_{1+}A_{T-1} + {}_{2+}A_{T+2} + {}_{3+}A_{T-3} + \ldots \quad \text{(equation 5)}$$

The four-day cycle policy involves a lot of work on one day and less on the other three. The staff might prefer to spread the load and each day change the sheet for a quarter of the rooms that REMAIN occupied; there are $G_{T-1} - D_T$ of these. Of course, the D_T beds vacated by the day's departures must also be changed. The appropriate model would be:

MODEL 4

$$SC_T = 2\left(D_T + \frac{G_{T-1} - D_T}{4}\right), \text{ every day}$$

which is quite a complicated equation for such simple-sounding behaviour. This attempt to 'smooth' the workload each day does not produce a constant daily number of sheet changes because both G_T and D_T vary.

Although sheets may be REMOVED from the bed of a guest on his departure, clean sheets may NOT be put on. This could be because the hotel is in a 'slack' period with few guests staying and they wished to avoid beds being left empty and becoming damp; another reason could be shortage of sheets. The model must be made slightly more sophisticated to express this.

Defining SON as the number of clean sheets going on to a bed, and *SOFF* as the dirty sheets removed from a bed, models 2A, 3A and 4A may be developed from models 2, 3 and 4.

Model 2 is the one where every guest's bed 'due' for a sheet change gets one, together with the day's vacated beds; the policy is the same for each day. The guests 'due' for sheet changes are those who have spent 4 and 8 nights in the hotel, ${}_{4+}A_{T-4} + {}_{8+}A_{T-8}$. For the amended model 2A, sheets will be taken off AND put on for these guests. In addition, dirty sheets will be removed from the beds of the guests who have departed D_T, and clean sheets put on for the new arrivals, A_T.

MODEL 2A

$$SON_T = 2(A_T + {}_{4+}A_{T-4} + {}_{8+}A_{T-8}) \text{ every day}$$
$$SOFF_T = 2(D_T + {}_{4+}A_{T-4} + {}_{8+}A_{T-8}) \text{ every day}$$

Model 3 may be amended similarly. This is the model where major sheet changes occur every fourth day and routine changes of newly-vacated beds take place in the intervening days. In the amended version of the model, 3A, the sheets taken OFF the beds are the same as before; the sheets put ON are different. On a routine day, clean sheets are put on only for the newly-arrived guests, A_T; on a sheet-changing day, clean sheets are put on for the guests remaining in the hotel and for the newly-arrived guests.

There are $G_{T-1} - D_T + A_T$ beds to be changed. If you think about this expression (and refer to equation 1) you can see that it refers to the guests who will be staying in the hotel tonight, G_T.

MODEL 3A

$$\left.\begin{aligned} SON_T &= 2G_T \\ SOFF_T &= 2G_{T-1} \end{aligned}\right\} \text{ if } T \text{ is a sheet-changing day}$$
$$\left.\begin{aligned} SON_T &= 2A_T \\ SOFF_T &= 2D_T \end{aligned}\right\} \text{ if } T \text{ is a routine day}$$

The amended version of model 4 is as follows:

MODEL 4A

$$SON_T = 2\left(A_T + \frac{G_{T-1} - D_T}{4}\right), \text{ every day}$$

$$SOFF_T = 2\left(D_T + \frac{G_{T-1} - D_T}{4}\right), \text{ every day}$$

This new sheet-changing policy spreads out the work involved in servicing guests who are already at the hotel ($G_{T-1} - D_T$ of them) while at the same time makes sure the new arrivals, A_T, get clean sheets and that the dirty sheets are removed from the beds of the D_T guests who have just left.

Once more, you can use your table of A's to identify the guests who will have their sheets changed under the policies modelled by 2A, 3A and 4A. The particular policy in operation depends on the establishment. Consider a 'bed-and-breakfast' guest house, a small seaside hotel and a large central London hotel and choose the model most appropriate for each.

The final models we shall consider are the simplest of all. Names are simply given to the sheets taken off and put on the beds without regard to what goes on in the hotel.

Black Box Model

MODEL 5

SC_T is the number of sheets changed each day

$SOFF_T$ is the number of sheets taken off the bed every day

The hotel is regarded as a 'black box' through which sheets pass. 'Black box' is a term used to describe a system that we do not want to look INTO; we are concerned only with what ENTERS or LEAVES the system.

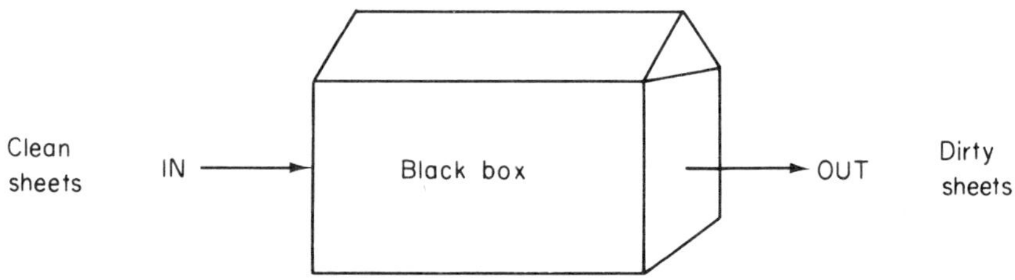

The selection of a model depends on three things:

(1) The problem being investigated.
(2) The accuracy of each model and the accuracy required.

(3) The cost and difficulty of obtaining the information necessary for each model; the reliability of such information.

Consider the first point. The reason we are interested in building a model is to establish how many sheets the hotel needs. Certainly the number of clean sheets required each day is relevant to the problem. The complete solution will involve other parts of the linen system: the laundry, the collection and delivery services, which must also be investigated.

By modelling the hotel part of the system in the ways considered we are 'opening up' this part and revealing its workings. No longer is 'hotel' a 'black box' through which sheets pass. We are investigating why and when the hotel will generate many or a few dirty sheets, and how it uses clean sheets. We do this by considering the two components of 'hotel': the guests and the domestic staff.

Instead of this:

we have this:

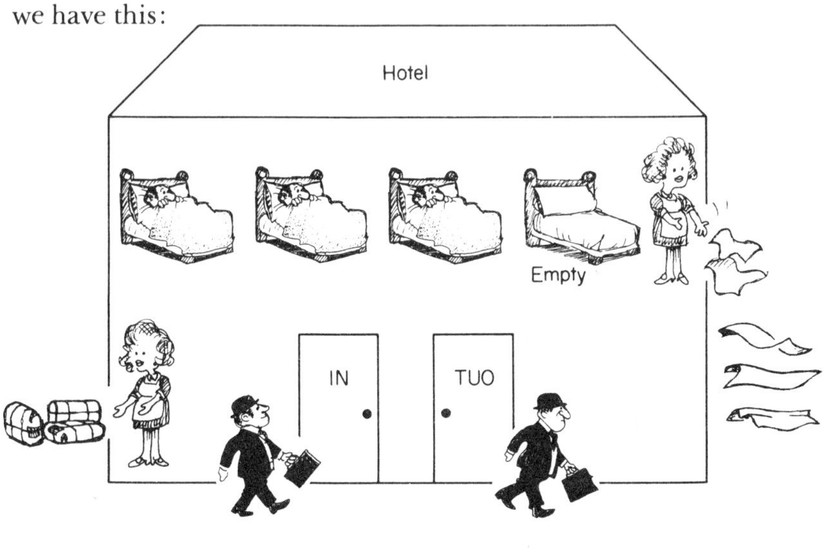

Unfortunately, opening a black box means more data must be collected. Broadly speaking the models 1–5 use data at three levels:

I. the $SOFF_T$, SON_T level;
II. the D_T, A_T, G_T level;
III. the ${}_{4+}A_T$, ${}_{8+}A_T$ etc. level.

Models using level I data are concerned only with the number of SHEETS going in and out of the hotel; where they go or how long they are on the beds is of no concern. These are the 'black box' models.

Models using level II data open the box a little and look at the ROOMS where sheets enter or leave. The model ignores who is in a room. Models using level III data go one stage further and investigate the behaviour of GUESTS.

Generally, the more elaborate the model the more detailed the information required. On this basis we can rank the models.

Model	Description	Data required on
5	$SOFF_T = SON_T$	$SOFF_T$
5A	SON_T $SOFF_T$	$SOFF_T$ SON_T
4	$SC_T = 2\left(D_T + \dfrac{G_{T-1} - D_T}{4}\right)$	D_T, G_T
4A	$SON_T = 2\left(A_T + \dfrac{G_{T-1} - D_T}{4}\right)$ $SOFF_T = 2\left(D_T + \dfrac{G_{T-1} - D_T}{4}\right)$	$D_T\ A_T\ G_T$
3	$SC_T = 2G_{T-1}$, sheet-changing day $SC_T = 2D_T$, routine day	$D_T\ G_T$
3A	$SON_T = 2G_T$, $SOFF_T = 2G_{T-1}$ } sheet-changing day $SON_T = 2A_T$, $SOFF_T = 2D_T$ } routine day	$D_T\ A_T\ G_T$
2	$SC_T = 2(D_T + {}_{4+}A_{T-4} + {}_{8+}A_{T-8})$	D_T, ${}_{4+}A_{T-4}$, ${}_{8+}A_{T-8}$
2A	$SON_T = 2(A_T + {}_{4+}A_{T-4} + {}_{8+}A_{T-8})$ $SOFF_T = 2(D_T + {}_{4+}A_{T-4} + {}_{8+}A_{T-8})$	D_T, A_T, ${}_{4+}A_{T-4}$, ${}_{8+}A_{T-8}$
1	$SC_T = 2(D_T + {}_{4+}A_{T-4} + {}_{8+}A_{T-8} + {}_{12+}A_{T-12} + {}_{16+}A_{T-16})$	D_T, ${}_{4+}A_{T-4}$, ${}_{8+}A_{T-8}$, ${}_{12+}A_{T-12}$, ${}_{16+}A_{T-16}$

(↓ Increasing complexity; ↑ Data more aggregate)

Data for the Models

As you go down the page the model becomes more elaborate and the information to be collected more detailed. Before we consider the various difficulties associated with collecting these sets of data, notice how lower-level data sets are aggregates of higher level. For example, in model 4A, SON_T, a level *I* variable, is made up of A_T, G_{T-1} and D_T which are level II variables. Similarly, the equations for D_T and G_T, both level II variables,

$$D_T = {}_1A_{T-1} + {}_2A_{T-2} + {}_3A_{T-3} + \ldots \qquad \text{(equation 3)}$$

$$G_T = {}_{0+}A_T + {}_{1+}A_{T-1} + {}_{2+}A_{T-2} + {}_{3+}A_{T-3} + \ldots \qquad \text{(equation 5)}$$

show that these are made up of the *A*'s, which are level III data.

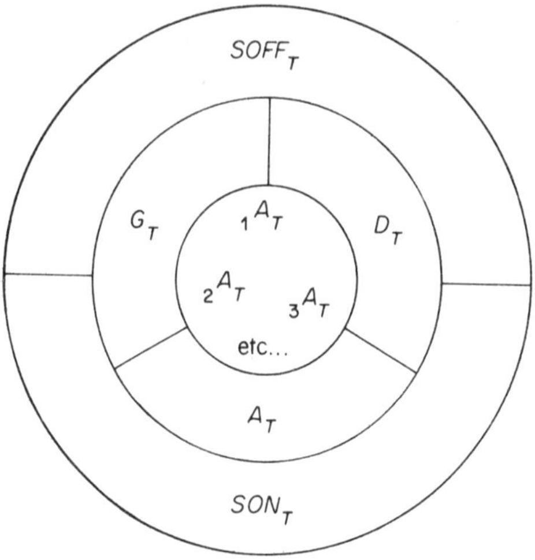

Levels of data

It is therefore possible to construct lower-level data sets from higher levels. Whatever the model and level of data selected, information needs to be collected for the different days of the week and weeks of the year in order to appreciate the variation in each variable under observation.

Consider the practical difficulties of obtaining each type of information. How would it be possible to determine $SOFF_T$ and SON_T, the sheets taken off and put on the beds? SON_T is the number of clean sheets taken from the clean-linen store. It is unlikely that daily stock-level figures would be kept and domestic staff would probably not want to be put to the inconvenience of counting the clean sheets removed from stock each day so SON_T may not be easily obtained. Similarly, staff would probably be even less happy if they were asked to count the dirty sheets, $SOFF_T$. So even though models 5 and 5A are simple they cause practical data-collection difficulties.

The next grade of model uses D_T, A_T and G_T, the number of guests leaving and arriving each day and the number staying each night. When a guest leaves he pays his bill so the cashier may have records of the guests leaving each day and arrangements could be made to collect the bills from her each day and use them as a data source. Presumably, when a guest arrives he checks in and his name and date of arrival are put in the register; information on A_T is therefore available. G_T, the number of guests staying each night, is more difficult. Leaving aside the possibility of prowling round the hotel each night and prodding all the beds, one way to find G_T would be from the register. This would only be possible if a record of each guest's departure was also kept. Generally, a hotel will keep a note of which rooms are let for the night in order that it can allocate a vacant room to a new arrival; this often takes the form of a chart of the room numbers and G_T could be calculated from this chart. Both these sources for G_T require someone on-the-spot to do some calculating EVERY DAY for as long as information is required. If this is possible then data is available for models 3, 3A, 4 and 4A.

The most complicated models require information on the numbers of guests arriving each day and the lengths of their visits. Examples of such data were given in Chapter 2 and in that chapter it was suggested that the source of such data should be the hotel bills. When a guest departs he (should!) pay his bill and on it would be a record of his arrival date and how long he stayed. If all these bills could be retained the required information could be taken from them. The advantage of this data source is that it already exists, no extra calculation or paper work needs to be done. Notice also (equations 3 and 5) that the other information on G_T, D_T, $SOFF_T$, SON_T could be obtained from it, enabling other models to be used. However, some calculation would be needed to do this and a more direct source may be preferred.

Each set of data is likely to have inaccuracies: mistakes may be made on the hotel bills; some bills may be lost. Effort should be made to estimate the extent of these errors.

Modelling the Linen System

Having selected a model and data source for the hotel, it is necessary to consider the rest of the system. We have assumed so far that clean sheets were always available should the hotel require them, without regard to how they got there. Similarly, the problem of the removal of dirty sheets was ignored.

If we assume that the laundry vans make enough trips each day from hotel to laundry to remove all the dirty sheets the hotel generates that day, and to deliver all the clean sheets processed by the laundry that day, then the vans do not interrupt the flow of sheets.

COLLECTION MODEL is
sheets collected from hotel_T = dirty sheets taken off the bed_T
$SCOL_T = SOFF_T$

DELIVERY MODEL is
sheets delivered to hotel on day_T = sheets processed by laundry_T
$SDEL_T = SWASH_T$

Of course, neither of these models may be appropriate. The vans may not make trips on Sundays for instance, so that the equations would apply only to the other days; on Sundays *SCOL* and *SDEL* would be zero. Furthermore, the number of dirty sheets generated on some days may be very high, for example Saturdays and Sundays when departures are high, and the vans may not make enough trips to carry them all. If the van could take 100 sheets per day the collection model would be:

$$SCOL_T \text{ is the smaller of } SOFF_T \text{ and } 100$$

which we write as:

$$SCOL_T = \text{minimum}\ (SOFF_T,\ 100)$$

and the delivery model would be:

$$SDEL_T = \text{minimum}\ (SWASH_T,\ 100)$$

Consequently the vans interrupt the flow of sheets round the system.

The remaining part to be modelled is the laundry and this can be done quite simply. If we assume that the laundry washes one day what it receives the day before then we have:

LAUNDRY MODEL

$$SWASH_{T+1} = SCOL_T$$

If there is an upper limit, say 70, on the number of sheets which can be washed then the model might be:

$$SWASH_{T+1} = \text{minimum}\ (SCOL_T,\ 70)$$

Other models, taking into account reduced or zero outputs on Saturday and Sunday or extra output using overtime, are also possible.

Now that the individual parts of the system have been modelled the pieces can be put together to see how they fit. You can imagine the system as a set of 'boxes' representing the hotel, laundry, etc., connected by string, and the sheets as beads on the string.

If you glanced at the system at any instant in time you would see the sheets in various places; as time passes the sheets travel round the system. It is a CLOSED system in that no sheets enter or leave it; an OPEN system would include the addition of new sheets and the

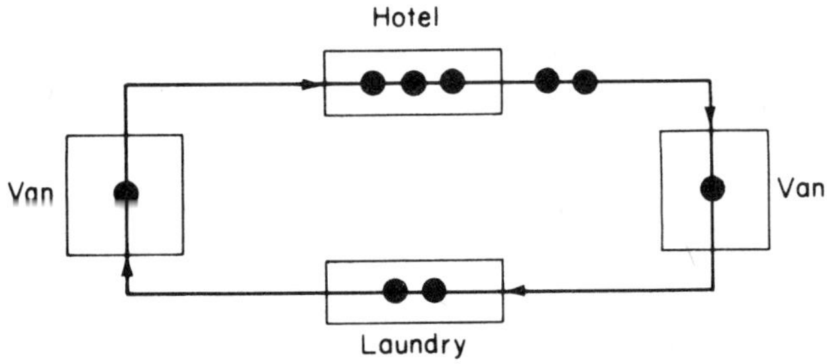

removal of worn-out or dirty sheets. The movement of the sheets is governed by the internal workings of the 'boxes': a high departure rate of guests from the hotel would cause many sheets to leave the hotel box; a machine breakdown at the laundry would increase the number of sheets waiting to be washed.

It is clear that the output from one box is the input for the next box in the cycle. But we have given mathematical names to the input and output of each box.

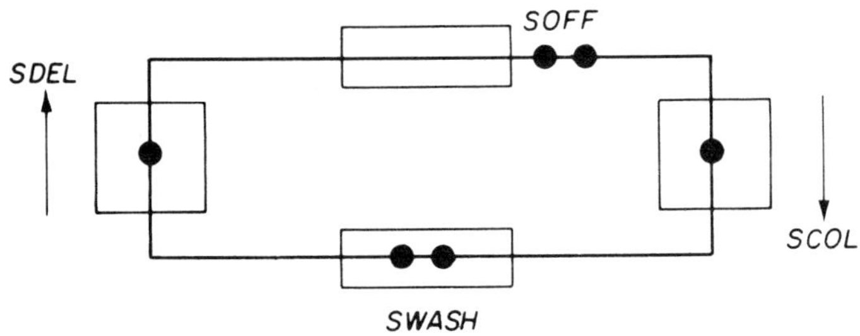

Moreover, we have in the models formulated mathematical expressions which describe the behaviour of these sheets. Let us follow the path of the dirty sheets taken off the beds on T-day. Using the simplest models for the laundry and vans, i.e. 'same-day' collection and delivery by the vans and previous day's washing done by the laundry, we have,

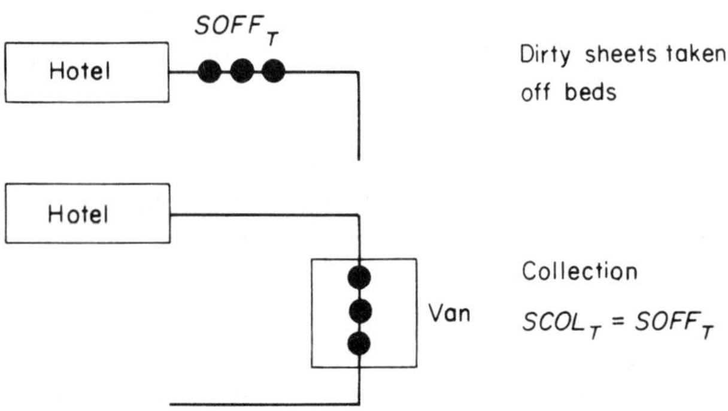

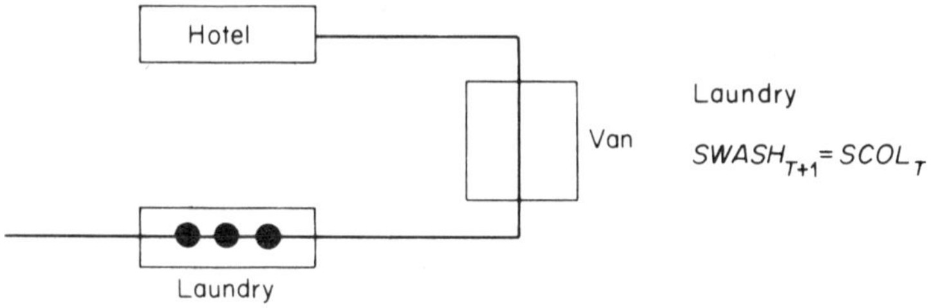

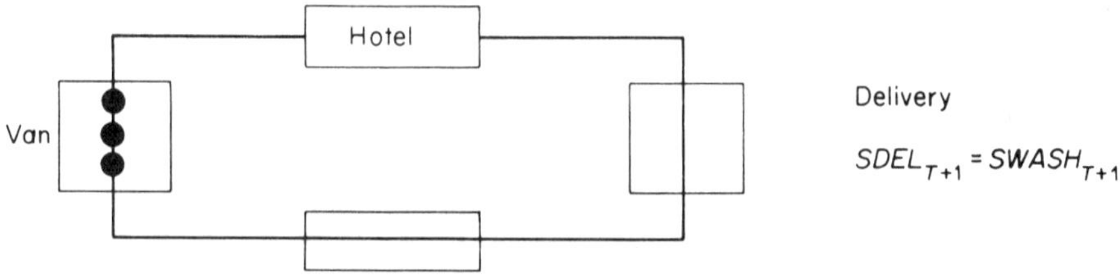

So the sheets which were taken off the beds on day T arrive at the hotel ready for use again on day $T + 1$, taking just 1 day to get round the rest of the system. We have built a model of the system which consists of the three equations:

$$SCOL_T = SOFF_T \qquad \text{COLLECTION}$$
$$SWASH_{T+1} = SCOL_T \qquad \text{LAUNDRY}$$
$$SDEL_{T+1} = SWASH_{T+1} \qquad \text{DELIVERY}$$

The model, though simple, has yielded a prediction. Its solution, $SDEL_{T+1} = SCOL_T$, has practical significance. If the hotel could be sure of sheets returning after 1 day from the laundry, this would go a long way towards determining how many sheets it should have. However, this is probably a rather optimistic picture and if one of the more realistic models were used, the time for the sheets to return would be rather longer. Whatever the actual time is, the appropriate models can be solved to find a good estimate of it. When the result is combined with the hotel part of the system, the hotel's sheets requirements can be found. This will be done in the next chapter (Model Analysis) for one set of models.

Modelling the Accommodation

An associated problem to the linen problem, and one that we have not considered so far, is the configuration of room sizes in the hotel: the number of double, single, twin and other sorts of rooms. The configuration obviously effects the number of guests who can stay and therefore the number and size of sheets to be laundered. Different configurations give different flexibility in allocating guests to rooms: double or twin-bedded rooms may be used for a single guest if all the single rooms are full; twin-bedded and double rooms may be interchangeable. Of course, each type of room takes up a different amount of space and if space is in short supply or expensive then the problem is to find the most economic set of rooms, subject to satisfying the guest's requirements.

Certain basic information is required: what are the guests' requirements and how much space does a certain type of room need? Many answers may be given to these questions. The size of a single room is a function of the furniture in it and BOTH depend on the level of luxury the hotel wants to offer and the type of clientèle it wants to attract. This is something that hotel management must decide. In order to proceed with the modelling and analysis of the problem we shall call the space taken up by a single room a 'space unit'; the space needed for a double and twin-bedded room is obviously more, maybe twice as much, 2 'space units' say.

A constraint which must obviously be satisfied is that the space used by all the rooms must be less than the space available for use. The number of rooms of each type will be limited by the available space.

One	single room uses	1	space units
Two	single rooms use	2	space units
Ten	single rooms use	10	space units
S	single rooms use	S	space units

Similarly, D, the number of double rooms, use $2D$ space units. Variables S and D are used for the number of single and double rooms because the actual values are not yet known. The total space used by all the rooms is:

$$S + 2D$$

The space available for bedrooms depends on the size of the hotel and the space needed for other facilities, e.g. kitchens, dining rooms, hallways etc.

Whatever the amount of available space, it must be true that,

Space used for for bedrooms	MUST NOT EXCEED	space available for bedrooms

If the amount of available space is B 'space-units' then,

$S + 2D$	IS LESS THAN OR EQUAL TO	B

There is (of course) a mathematical symbol for 'LESS THAN OR EQUAL TO'. Using it, we get:

$$S + 2D \leqslant B$$

This expression is called a CONSTRAINT; it shows how the values of S and D are constrained. Mathematically, the name for the expression is an 'inequality' because it uses the 'inequality' sign, $\leqslant$ (rather than the 'equality' sign, $=$). You may not have thought about it in this way before, but an equation is also a constraint. An equation involves one or more variables and before it is written those variables are 'free', they can take any value they like; but as soon as you write the equation down, you have constrained them. For example, the equation:

$$S + 2D = B$$

is a constraint. It says 'all the space available for bedrooms must be used for bedrooms, none must be left over'. This is quite different from the inequality above which allows left-over space AS WELL AS the case where all the space is used up (space used is less than OR equal to B).

You have seen equations relating two variables before (in Chapter 3), and you should be able to plot the equation on a graph. Which variable should be put on the vertical axis, which on the horizontal axis? The answer is, it doesn't matter. In Chapter 3 an equation showed the dependence of one variable on another, and generally the dependent variable is plotted on the vertical axis. However, the equation considered here involves two variables of equal status. We shall put S on the vertical axis. You can put D, and reconstruct the whole of the following argument: rough sketches will do. If the space available is 1000 space units, the graph looks like the diagram over the page.

The easiest way to construct the line is to find the points U and L where the line crosses the axes.

At U, $D = 0$, therefore:

$$S + 2D = 1000 \text{ becomes } 2D = 1000, \text{ so } D = 500$$

At L, $S = 0$, therefore:

$$S + 2D = 1000 \text{ becomes } 2D = 1000, \text{ So } D = 500$$

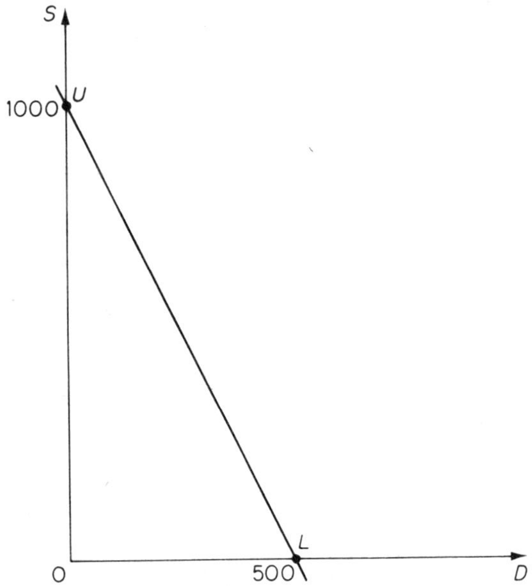

The line passes through the points (0, 1000) and (500, 0). Any point on the line represents a value for *S* and a value for *D*. For example, point *M* represents $S = 500$ and $D = 250$.

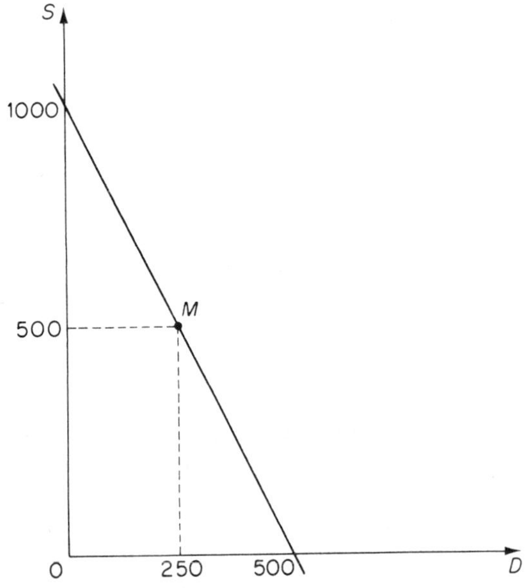

Moreover, the values of *S* and *D* represented by any point on the line satisfy the equation:

$$S + 2D = 1000$$

For example, if $S = 500$ and $D = 250$,

$$S + 2D = 500 + (2 \times 250)$$
$$= 1000$$

So the line 'selects' the points on the graph which represent values of S and D that satisfy the equation. Of course, some points have no practical relevance, for example point N, where $S = -200$ and $D = 600$.

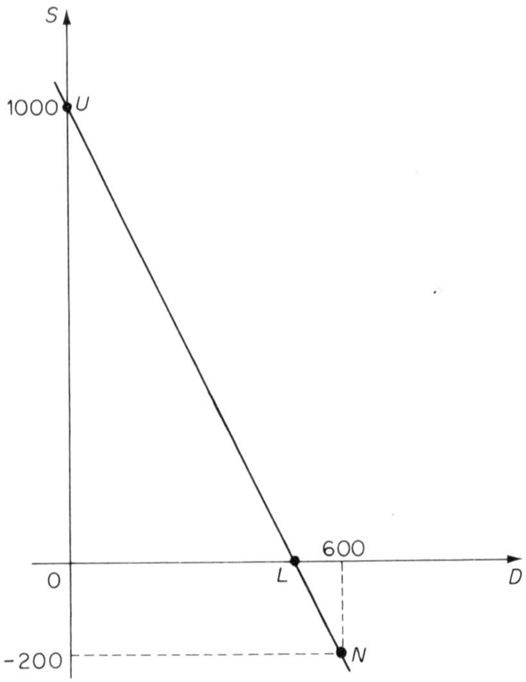

A negative number of rooms is meaningless. When we consider the values of S and D that satisfy the equation:

$$S + 2D = 1000,$$

we wish to consider only positive values.

S must not be negative
D must not be negative

These can be represented mathematically by constraints,

$$S \geqslant 0$$
$$D \geqslant 0$$

Then the points on the graph which represent values of S and D that are allowed by both the equation and the two new constraints lie on the section UL of the line. We call these points FEASIBLE points, because they represent possible solutions. So much for the equation, $S + 2D = 1000$, what about the inequality

$$S + 2D \leqslant 1000?$$

Can the graph show us the points representing values of S and D allowed by this? We already know some of these points, those on the line UL, but there are obviously others. For example, point (0, 0) is a feasible but not very interesting solution. Other feasible points are (200, 200), (300, 200), (100, 400) and there are a great many more. If you plot them you will find they all lie 'to the left of' the line, $S + 2D = 1000$.

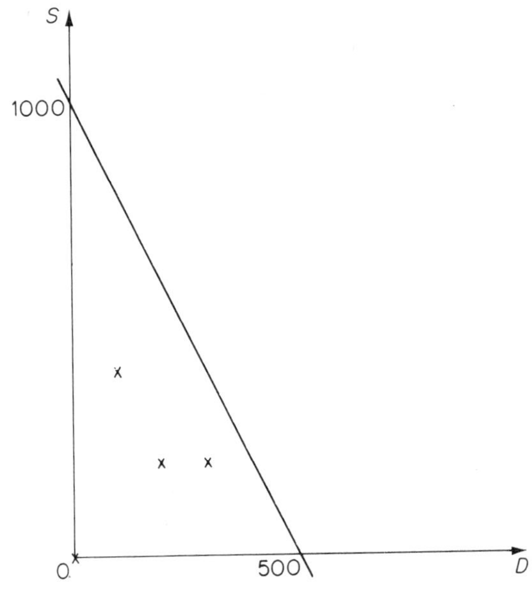

ANY point to the left of the line satisfies the inequality

$$S + 2D \leqslant 1000$$

Moreover, any point to the right of the line (for example, $D = 500$, $S = 500$) does NOT satisfy it. Try some and see. So the line conveniently divides the points into two groups; those that satisfy the constraint and those that do not. The 'infeasible' side of the line is shaded to show this.

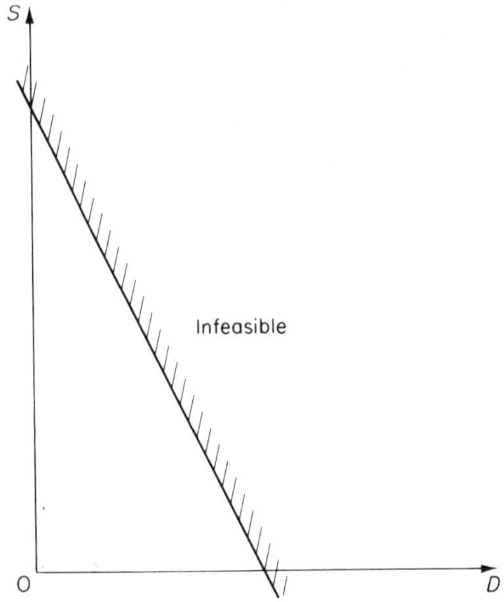

The points allowed by the other constraints

$$D \geqslant 0$$
$$S \geqslant 0$$

can be similarly shown.

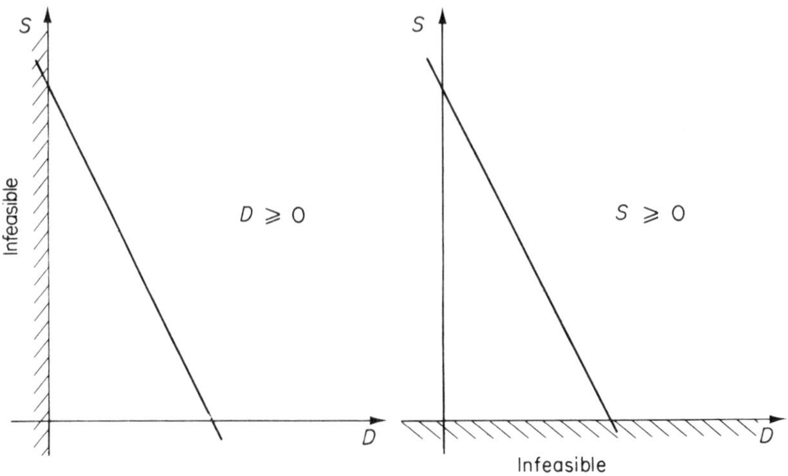

Notice how the line which divides the feasible and infeasible points for the constraint, $D \geqslant 0$, is the line $D = 0$, the S-axis. Similarly, the D-axis divides the points for the constraint $S \geqslant 0$.

When all the infeasible areas are marked on the same graph, an area remains which contains points which satisfy ALL the constraints

$$S \geqslant 0$$
$$D \geqslant 0$$
$$S + 2D \geqslant 0$$

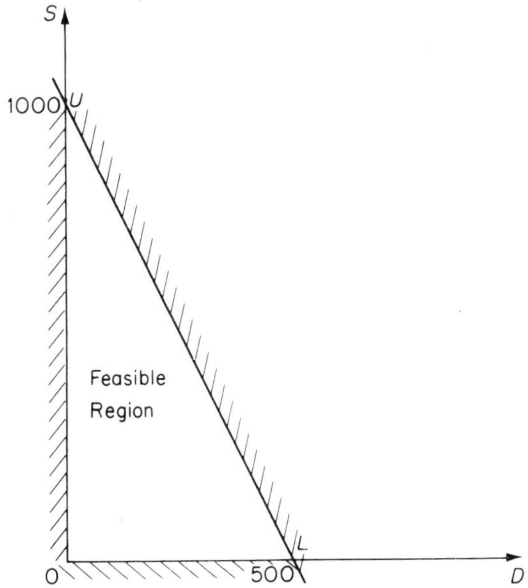

This area is called the feasible region and is represented by the triangle $U0L$. The feasible region for the constraints

$$D \geqslant 0$$
$$S \geqslant 0$$
$$S + 2D \quad \text{EQUALS} \quad 1000$$

was the line UL only, so the graph demonstrates what we discussed before, that an inequality is less constraining than an equation.

We now have before us all the points representing values of S and D, the numbers of single and double rooms that do not exceed the space available. At the moment, solutions with no double rooms are allowed: for example, $D = 0$, $S = 1000$. But management may not want to provide only single rooms and may prefer to offer a selection that meets guests' requirements; indeed, there might not be enough demand to fill 1000 single rooms. Other constraints can be formulated to represent these restrictions. If, from past behaviour, it was thought unlikely that there would be demand for more than 300 double and 800 single rooms, we can say

$$D \leqslant 300$$
$$S \leqslant 800$$

The graphs show the feasible and infeasible solutions.

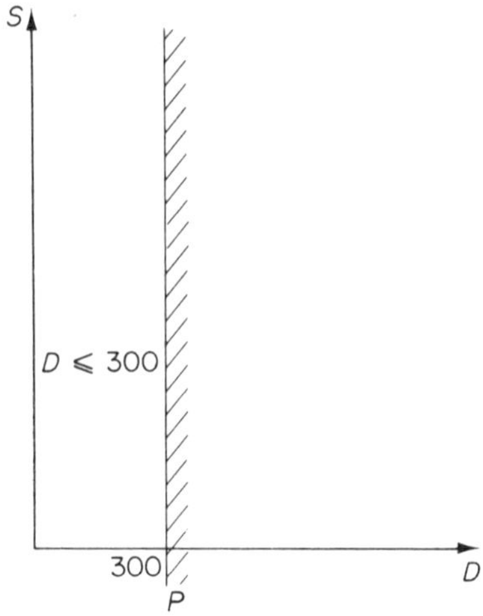

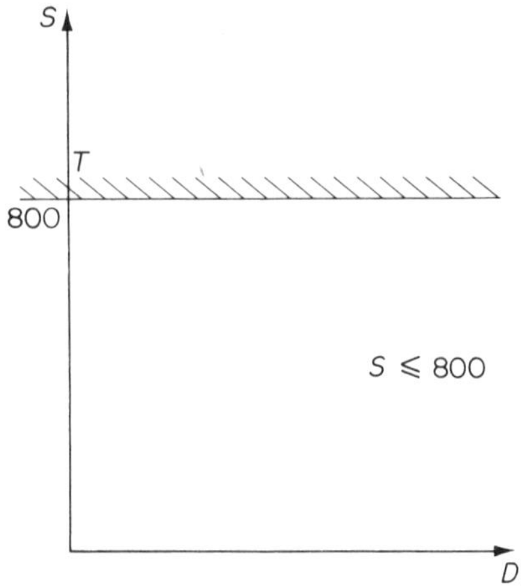

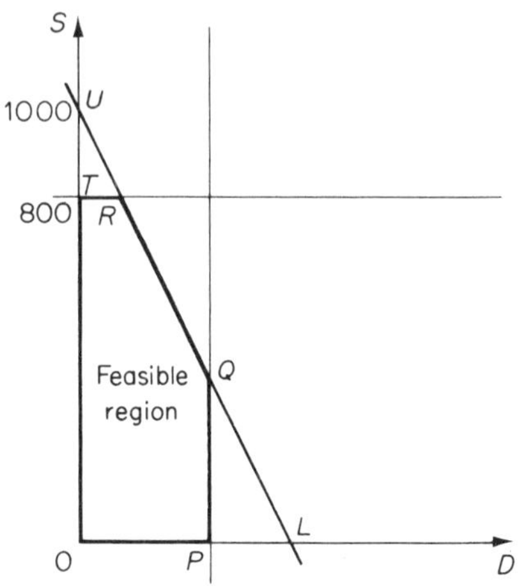

$$D \leqslant 300$$
$$S \leqslant 800$$
$$S + 2D \leqslant 1000$$
$$S \geqslant 0$$
$$D \geqslant 0$$

The AREA defined by o*PQRT* is the feasible region. It contains points which are allowed by the space constraint AND the two demand constraints. Notice how each of the demand constraints has 'chopped off' a portion of the original feasible region o*UL*, and thereby excluded solutions which WERE feasible.

Other models may be formulated for other situations. For example, the provision of a minimum number (say 50) of double rooms could be built in by adding the constraint

$$D \geqslant 50$$

This constraint does not allow solutions 'to the left of' the line *VW*. The amended feasible region is then *WPQRV*.

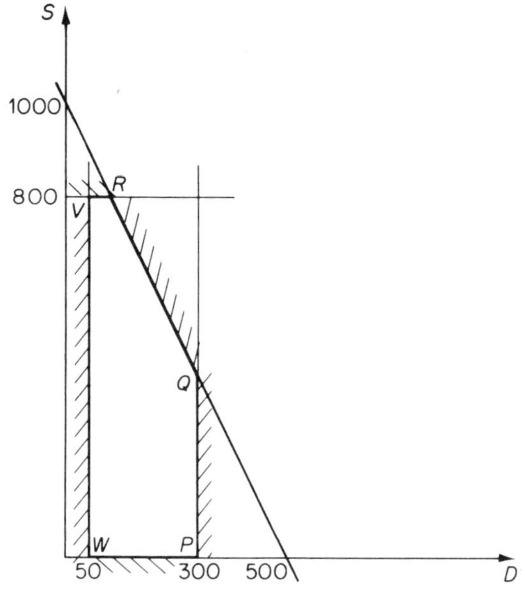

A further constraint might be

$$S \geqslant 2D$$

Find an interpretation for this constraint and find the new feasible region.

Solving the Accommodation Model

When the appropriate model has been developed, several solutions are possible, each one inside the feasible region. Which solution should be chosen? Clearly each represents quite different facilities in practice and some yardstick is needed to judge the merits of one against another. Presumably the hotel is in business to make money and it may want to offer the configuration of rooms that will bring in the greatest revenue. The price of an overnight stay is set by the hotel.

If this is £8 for a double room and £5 for a single room then IF ALL THE ROOMS ARE FULL the revenues received will be:

$$D \text{ rooms @ £8 each} = £8D$$
$$S \text{ rooms @ £5 each} = £5S$$
$$\text{Total revenue} = £8D + £5S$$

The revenue depends on the values of S and D. Clearly some solutions are better than others. For example $D = 200$, $S = 200$, gives a revenue of £2600, while $D = 100$, $S = 600$, gives a revenue of £3800.

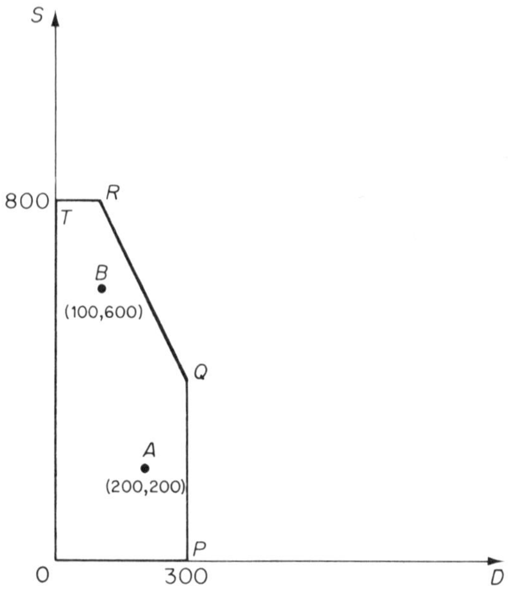

It is useful to identify those points on the graph which yield the SAME revenue. For example, point C ($D = 100$, $S = 360$) yields a revenue of £2600, the same as point A. Other points on the graph will also represent solutions giving revenue of £2600.

For all these solutions it must be true that

$$8D + 5S = 2600$$

But this is the equation of a straight line. It can be written as

$$S = \frac{2600}{5} - \frac{8}{5}D$$

The intercept is 2600/5 and the slope is –8/5.

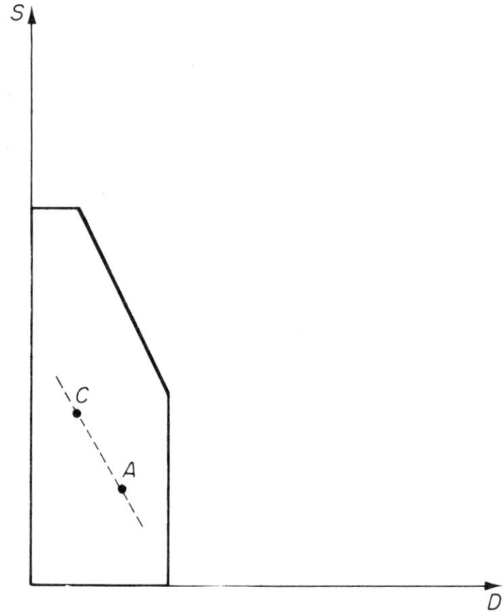

All the solutions giving a revenue of £2600 lie on the dotted line through *AC*. Similarly, all those giving a revenue of £3800 lie on the dotted line through *B*. These two lines are parallel (why?).

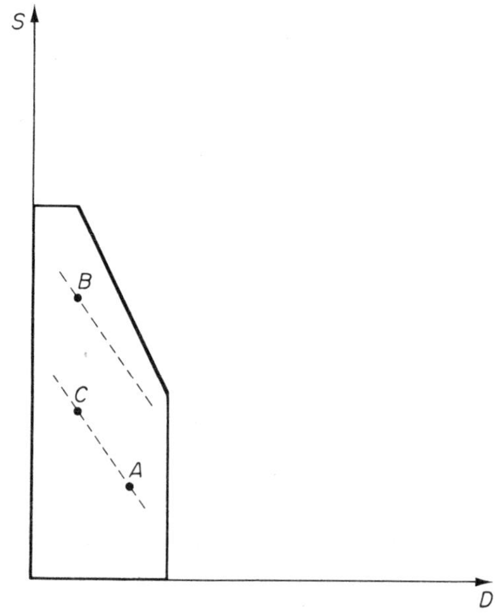

Solutions giving higher revenues lie on lines parallel to these two and to the 'right' of them. Some lines are completely outside the feasible region and represent revenues that are not attainable, for example line ④ below.

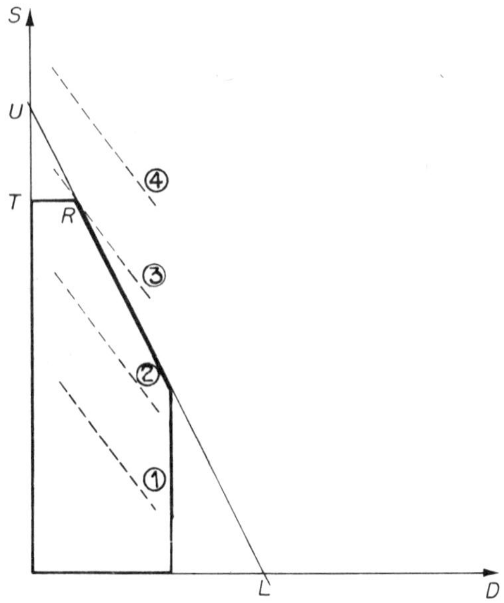

The best solution giving the largest revenue is point R; this is the only feasible point on the revenue line ③. Any line of higher revenue contains only infeasible points; any line to the left of ③ represents less revenue.

R is the intersection of lines TR and UL.

The line TR is represented by $S = 800$

$$UL \text{ is represented by } S + 2D = 1000$$

Therefore the values of S and D can be found by solving these two equations.

$$S = 800$$
$$\text{and } D = 100$$

$$\text{revenue} = 8D + 5S = £800 + £4000$$
$$= £4800$$

This is the maximum that can be attained subject to the space and demand constraints. (This way of treating the accommodation problem, formulating constraints and a function to be maximised, has the structure of the LINEAR PROGRAMMING approach to problem solving. For more information consult P. G. Moore's *Basic Operational Research.*)

The problem for the hotel is therefore solved: it should have 100 double rooms, 800 single rooms and it will receive £4800 each night—or will it? Suppose some rooms remain empty? All rooms offered are not necessarily filled. This would only be true in the case of excess of demand over supply, for example at an airport hotel when passengers are delayed or a seaside hotel at the height of the season. Even though the hotel can hope to achieve this maximum revenue at the busy periods of the year there is no guarantee that it will do so all the time.

It should be clear that this model relies very heavily on collected data. The piece of information insisting there should be no more than 800 single rooms is very critical. The final solution, and therefore the prediction, would have been radically different if this had been either 600 or 1200 for instance.

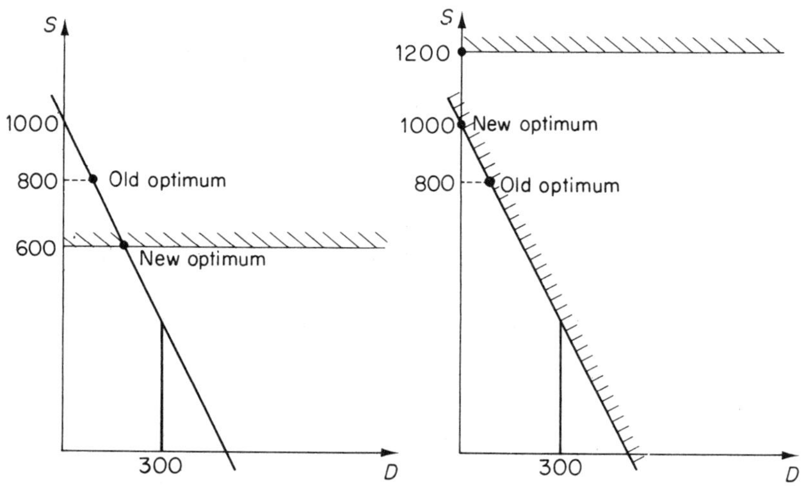

In the first case, the constraint $S \leqslant 600$ is more restrictive than $S \leqslant 800$, the new optimum is $S = 600$, $D = 200$, and the maximum revenue that could be achieved is £4600. In the second case, the constraint $S \leqslant 1200$ is REDUNDANT. Another constraint, the space constraint, is dominant and the new optimum is $S = 1000$, $D = 0$ with a maximum revenue of £5000.

The constraint which describes the demand for double rooms is not quite so critical. Putting $D \leqslant 200$ or $D \leqslant 400$ in the original model makes no difference to the optimal solution. Try it and see. The solution is still at R but the feasible region is different. What happens if $D \leqslant 50$?

Whatever the correct upper limits are on the demand for single and double rooms, the figures used in the model are estimates of these and it is a very worthwhile exercise to 'see what happens' to the solution, if these were varied. Even if the estimates used are good ones they may not be quite so good in 5 years' time if the type of clientèle changes. For example, at present it may consist almost entirely of businessmen who require single rooms. The attractions of the town in which the hotel is situated may increase over the next few years and the hotel's clientèle may then include a substantial element of tourists with an increased demand for double rooms.

If management feel there is more potential demand than they are catering for, then they must consider providing more space, i.e. relaxing the space constraint. For example, if they wish to provide 200 double AND 800 single rooms, represented by point A on the diagram, this is infeasible and the space provided must be increased to at least 1200 units.

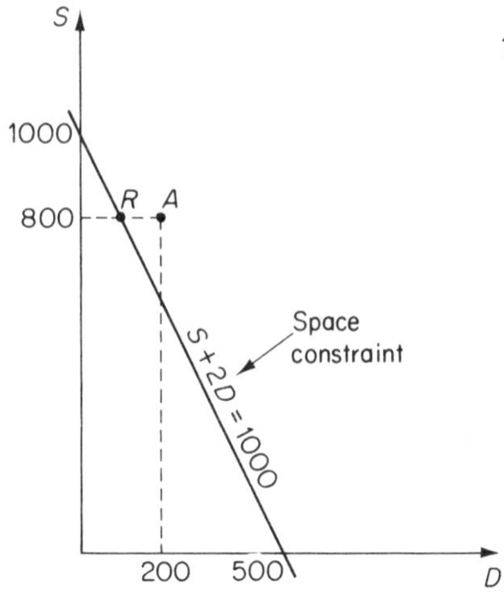

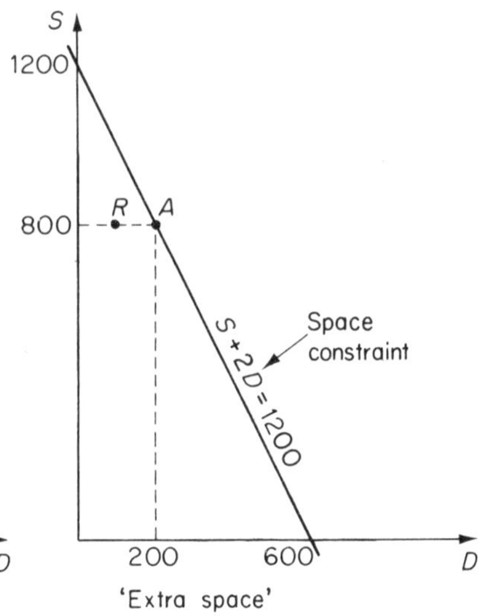

In addition to using the model to investigate the effect of different constraints, we can also determine the implications of varying the room charges. For example, if the price of a double room were increased to £11, the equation for total revenue becomes

$$\text{Revenue} = 11D + 5S$$

The feasible region remains unchanged because the constraints are the same.

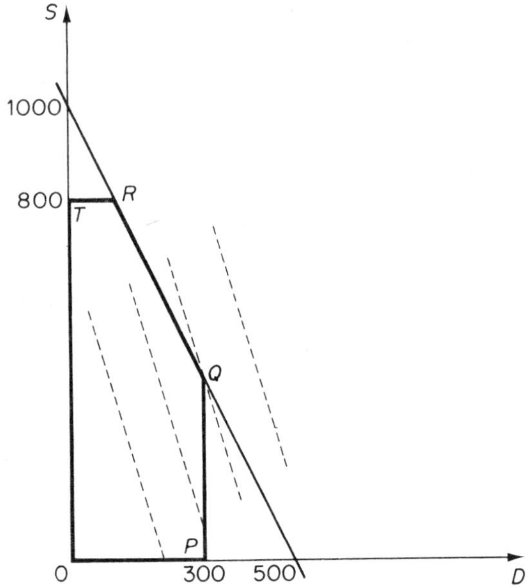

The dotted lines on the diagram connect solutions of equal revenue and have a greater slope than those of the previous revenue function. The optimum (maximum revenue) solution is now at Q, NOT at R.

$$D = 300$$
$$S = 400$$

and the revenue achieved is £5300. This is a result that could NOT have been found by simply inserting the previous optimum values of S and D into the new revenue function. We need the model to demonstrate that the very nature of the solution is different under the new pricing structure. As with the first revenue function, there is no guarantee that the maximum revenue will be obtained. Management may have seriously overestimated demand for double rooms for instance, and may have to let some of the 300 as singles at a loss. Indeed, demand may change as a direct result of a new price structure. Then the constraints are wrong and so is the solution of the model. The results of any model are only as good as the assumptions on which it is based: no matter how clever the mathematics used in finding a solution, if the assumptions are not valid, the model is not valid and the solutions derived from it will be meaningless.

6
Model Analysis

Model Assembly and Total Error

In the previous chapter various models of each part of the linen system were developed and a model of the SYSTEM was constructed by selecting the simplest of these for its components.

SYSTEM MODEL

$$\text{COLLECTION}: SCOL_T = SOFF_T$$
$$\text{LAUNDRY}: SWASH_{T+1} = SCOL_T$$
$$\text{DELIVERY}: SDEL_{T+1} = SWASH_{T+1}$$

Now we must ask ourselves 'How good a model is this?' Any system model must be examined from three viewpoints:

(1) How difficult it is to solve.
(2) Its accuracy.
(3) The difficulty of collecting the data it needs.

The overriding criterion is accuracy. The model is a tool, used to solve the problem of how many sheets the hotel needs, and we must decide on the overall desired level of accuracy of the results it gives. For example, we might require that the model should predict the hotel's sheet requirements to within ± 50 of the real value. We shall call this inaccuracy (± 50) the TOTAL ERROR of the system model.

The system model is made up of model components, each of which represents the behaviour of some part of the system and predicts how many sheets will leave that part.

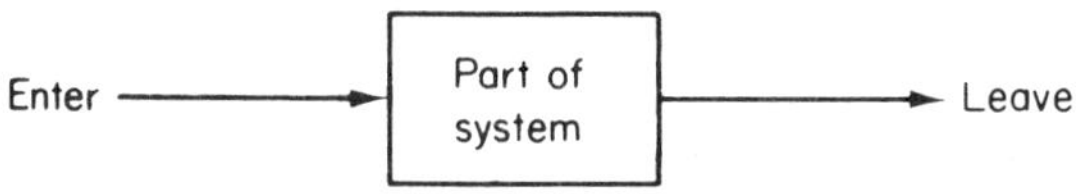

There is an error associated with this prediction, and the TOTAL ERROR is the accumulation of these separate errors. Therefore, the size of the total error imposes limitations on the size of the separate errors of the model components; the error of each component depends on the model and data used. The model components chosen must be sophisticated enough to meet the accuracy restrictions, and the components in turn demand data of sufficient detail and reliability, in keeping with their level of sophistication. A model runs on data in much the same way as a car runs on petrol and you don't put paraffin into a Rolls-Royce . . .

It is clear then that the desired level of accuracy determines what models to use and what data to collect. Ideally, all components will have about the same error. Some may be more elaborate than others because the behaviour they represent may be more complicated; similarly, some will need a lot of detailed information. However, this is not the case for the system model constructed earlier. All the components are equally elaborate but not equally ACCURATE. Assembling a system model from models of varying accuracy makes about as much sense as making a Hi-Fi system from a superb deck and amplifier and using cardboard boxes for speakers: the resulting system is only as good as its worst component. It is better to decide on what system you can afford (or require) and then choose the pieces accordingly; exceeding the requirements only leads to unnecessary expense. There are two areas of expenditure when using models to solve problems: collecting the data and solving the model. Ideally, one would like to have a model that was accurate, easy to solve and relied on data that was easy and cheap to collect but, unfortunately, these aims are conflicting. Higher accuracy usually implies a more elaborate model and more detailed data; the elaborate model requires longer to solve whether this is done manually or on a computer, and the detailed data is more expensive, in time and manpower, to collect. Every manager should pay particular attention to these two areas of expense whether he is conducting the study himself or using a team of specialists. He should let the total error dictate the type of model and data required and not waste either time or money exceeding the requirements, e.g. collecting too much data. Manager control is by total error control.

The crux of the matter of model selection is to determine the accuracy of a particular model component in representing the practical situation. This depends on the variability of the practical situation and, as we discussed in Chapter 2, the only way of establishing the variability of the situation is to collect some data on it.

The collection, delivery and laundry sections probably do not vary much at all because they are almost entirely under managerial control; their behaviour is regulated. The number of sheets washed at the laundry, for example, depends on the machine capacity and the labour force, both of which are relatively constant. It also depends on a managerial decision on how these resources are used. Only an eccentric would arrange for a laundry to wash 200 sheets one day, 10 the next, 50 the next and so on, so the laundry activity should be quite simple to model. There are factors outside the system 'called 'exogenous' factors) which might affect laundry output: there may be labour unrest, from time to time; the machines may break down. But, hopefully, these exogenous factors will not play both a significant AND continuing role.

The part of the system exhibiting the most variability is probably the hotel because so many of the contributing factors are outside managerial control. Although the hotel manager can control the number of sheet changes a guest has DURING his stay, he cannot affect those necessary on the arrival or departure of guests; neither is he in complete control of the NUMBER of guests staying. All this depends on the guests themselves. Guests come to stay at a hotel for all kinds of reasons and there is no reason to believe that they will behave in the regular way we would like with, for example, exactly 20 people arriving every day and each staying one night. When people have a wide variety of choice they generally take advantage of it; their behaviour is not under the control of the system. So quite a sophisticated model is probably needed to represent accurately the full variations of the number of guests staying and the length of their visits. Such models were developed in Chapter 2. Unfortunately, as we have already stressed, sophisticated models are often lengthy and difficult to solve so it will not be possible in this book to proceed further with the analysis of the linen system using these models. Instead we shall reduce and simplify them. Later on we shall discuss the treatment of the original models.

Simplifying the Linen Problem

The hotel is a large complex organisation catering for a wide variety of clientèle. We can simplify the hotel part of the system by considering instead a small bed-and-breakfast guest house.

First the simplifications:

(1) We will make do with one sort of behaviour, which will apply to guests arriving on any day of the year.
(2) Each guest stays only one night.

Now, the reductions:

(1) Our hotel can accommodate only 4 guests at a time.
(2) Not more than 6 guests arrive on any one day, hoping to be accommodated.

Once the variations of the day of the week and the length of visit have been made constant, the only variation left is in the number of guests arriving each day.

The following pattern of behaviour will be used:

x = number of guests wishing to stay	0	1	2	3	4	5	6
$p(x)$ = probability of x guests arriving in a day	0·04	0·02	0·04	0·20	0·35	0·20	0·15

Such a pattern could have been constructed from collecting data on the number of guests arriving each day and calculating the proportions of zeros, ones, twos, etc. The proportions are used for probabilities in the model. This assumes that the same pattern of behaviour will persist. The row of probabilities contain the 'essence' of the pattern, showing how likely certain values of x are. For instance, very rarely (chance of 0·04) would no guests arrive at the hotel; put another way, you would expect to have no arrivals on 4 days out of every 100, on the average. Most likely (chance of 0·9) there would be 3 or more arrivals. What is the mean number of arrivals?

In 100 days, you would expect:

4 days with no arrivals
2 days with 1 arrival
4 days with 2 arrivals
20 days with 3 arrivals
35 days with 4 arrivals
20 days with 5 arrivals
15 days with 6 arrivals

The mean number of arrivals

$$= \frac{\text{total number of arrivals in 100 days}}{100}$$

$$= \frac{(4 \times 0) + (2 \times 1) + (4 \times 2) + (20 \times 3) + (35 \times 4) + (20 \times 5) + (15 \times 6)}{100}$$

= 4 arrivals per day

Using a technique called 'simulation' it is possible to generate groups of guests who will arrive according to the probability pattern and with a mean of about 4.

Simulating Arrivals

Consider a wheel (and a pointer).

The wheel is divided into numbered sections and the proportion of

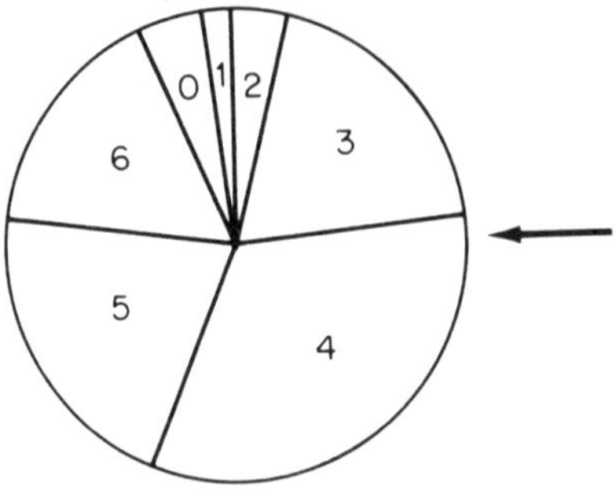

the wheel allocated to a particular number is equal to the probability of that number of guests arriving on a day.

If the wheel is spun freely about its centre, it will eventually stop with one of the numbers opposite the pointer. It is clear that if the wheel is spun many times, some numbers will come up more often than others. For example, the proportion of threes will be 0·20, the proportion of ones will be 0·02. In fact, the numbers obtained will follow the probability pattern. Make a paper disc the size of a record and trace on it the numbered sections from the diagram above. Spin the disc 100 times on the record player and write down the numbers obtained. Use a fixed spot on the player as the pointer. Calculate the proportions and the mean.

Below are 20 numbers presented in the order in which they were obtained, from 20 spins of the wheel.

3	3	4	3	3	3	4	5	4	4
0	3	6	3	4	4	6	4	4	3

The frequencies and proportions are:

x	0	1	2	3	4	5	6
Freq.	1	0	0	8	8	1	2
Prop.	0·05	0	0	0·4	0·4	0·05	0·1

Compare this with the original probability pattern. There is some discrepancy certainly, but this is to be expected when only 20 numbers are considered. Two 'extra' threes rather than fives distorts the picture. Your 100 numbers should provide a closer matching.

The mean is 3·65. 'Not very like 4' you might say, but nonetheless the numbers do roughly follow the pattern and they are a set of figures that COULD have been collected by counting the number of guests arriving on each of 20 successive days – and that is precisely the interpretation we shall now give them.

The 20 numbers simulate the arrival of guests at the hotel. Using the reduced hotel model which will accept no more than 4 guests each night, we can see how these 'arrivals' will be treated and what their effect will be on the hotel. We can determine how many guests will be turned away because the hotel is full, and how many rooms will be vacant each night.

Consider the 3 guests who arrive on the first day: they will all be accommodated and there will be one vacant room. On the second day exactly the same thing happens. On the eighth day 5 guests arrive so 1 will be turned away and no rooms are vacant. The following table shows what happens on each of the 20 days.

Hotel simulation

Day	*Arrivals*	*Number staying*	*Number turned away*	*Number of vacant rooms*
1	3	3	0	1
2	3	3	0	1
3	4	4	0	0
4	3	3	0	1
5	3	3	0	1
6	3	3	0	1
7	4	4	0	0
8	5	4	1	0
9	4	4	0	0
10	4	4	0	0
11	0	0	0	4
12	3	3	0	1
13	6	4	2	0
14	3	3	0	1
15	4	4	0	0
16	4	4	0	0
17	6	4	2	0
18	4	4	0	0
19	4	4	0	0
20	3	3	0	1

As the last two columns show, MOST of the time all the rooms are full and all customers are satisfied. The non-zero entries in these columns represent discrepancies between supply and demand and may indicate areas for improvement. Altogether 73 guests arrive in the 20 days and 5 of these (7 per cent of total) are turned away. This represents lost revenue to the hotel, for these guests do not 'queue' for accommodation: once turned away they stay away. Similarly, in the 20-day period the hotel has available 80 'nights worth' of accommodation; 12 of these (15 per cent of the total) remain unfilled representing unused capacity.

If the simulation was carried on for hundreds of days, the mean number of 'arrivals' would be very close to 4. The total of unfilled rooms and the total of unsatisfied guests would be equal because the mean demand equals the mean supply. However, by catering for the mean the hotel would still be losing custom because the varied arrival pattern of guests DOES NOT MATCH the constant supply patterns of rooms. For complete matching, either the guests must arrive 4 every day – which is beyond the hotel's control – or the rooms available must match the irregular arrival of guests (a clearly absurd suggestion). Faced with this dilemma the hotel must consider offering more rooms than that required by the AVERAGE number of guests arriving but exactly how MANY more depends, of course, on the 'cost' in terms of capital tied up. Management must weigh this cost against the extra revenues it might hope to achieve by having more rooms available.

Now that the hotel part of the system has been copied, the results in terms of dirty sheets can be fed into the next part, collection. The number of (pairs of) dirty sheets generated TOMORROW is equal to the number of guests staying TODAY. The average number of dirty sheets over the 20-day period is 3·4 per day which is less than the average number of arrivals (3·65) because some were turned away. Now, in order to illustrate a point, let us assume that the laundry can handle 4 pairs of sheets every day except Sunday, when it is closed. You can confirm that the mean capacity over 7 days is about 3·4 per day, equal to the mean arrival of dirty sheets. Let us see what happens when this 'mean-matched' system operates.

'Mean-matched' Systems

If day 1 is a Monday, then the first batch of dirty sheets are removed from the beds on Tuesday. Collection ensures they reach the laundry some time the same day.

Collection simulation

	Day	*Guests staying*	*Dirty sheets at hotel*	*Arrival of sheets at laundry*
Mon	1	3	–	–
Tues	2	3	3	3
Wed	3	4	3	3
	–	–	–	–
	–	–	–	–
	–	–	–	–

On subsequent days the process is similar, except of course that no sheets 'arrive' on Sundays. These stay at the hotel until Monday when they are collected with that day's dirty sheets.

Collection simulation

	Day	Guests staying	Dirty sheets at hotel	Arrival of sheets at laundry
Mon	1	3	–	–
Tues	2	3	3	3
Wed	3	4	3	3
Thurs	4	3	4	4
Fri	5	3	3	3
Sat	6	3	3	3
Sun	7	4	3	0
Mon	8	5	7	7

The sheets are washed the day after they reach the laundry, except for sheets which arrive on Saturday which must wait until Monday.

Laundry simulation

	Day	Arrival of sheets	Sheets queueing for wash	Sheets washed	Sheets not washed	Spare capacity number of sheets
Tues	2	3	—	—	—	—
Wed	3	3	3	3	0	1
Thurs	4	4	3	3	0	1
Fri	5	3	4	4	0	0
Sat	6	3	3	3	0	1
Sun	7	0	3	0	3	0
Mon	8	7	3	3	0	1

If more than 4 sheets are queueing to be washed, the excess must remain in the queue to be joined by the next day's arrivals.

	Day	Arrival of sheets	Sheets queueing for wash	Sheets washed	Sheets not washed	Spare capacity number of sheets
Mon	8	7	3	3	0	1
Tues	9	4	7	4	3	0
Wed	10	4	7	4	3	0

The table continues as follows:

Tues	9	4	7	4	3	0
Wed	10	4	7	4	3	0
Thurs	11	4	7	4	3	0
Fri	12	0	7	4	3	0
Sat	13	3	3	3	0	1
Sun	14	0	3	0	3	0
Mon	15	7	3	3	0	1
	—	—	—	—	—	—
	—	—	—	—	—	—
	—	—	—	—	—	—

Complete the table for yourself. You will find that there are always more sheets queueing to be washed than the laundry can handle. The sheets wait longer and longer in the queue and the last sheets of the 20-day period, which leave the hotel on the 21st day, are not washed until the 24th day. If the simulation were carried on for a longer period the queue would build up even more and the laundry would become a bottleneck in the system. This is so even though the mean arrival of sheets and the mean capacity of the laundry are the same. It is caused by the difference in the patterns of arrival and of washing capacity. To illustrate: the pattern of arrival for the first 7 days is:

Tues
3 3 4 3 3 0 7, 23 sheets in total

and the pattern of washing capacity is:

Wed
4 4 4 4 0 4 4, 24 sheets in total

Even though the washing capacity over the 7 days is about the same as the arrivals, some capacity is 'lost' in the first few days when capacity exceeds the dirty sheets, and this leads to a shortage later on and, of course, queueing. The lesson to be learnt is that 'mean-matched' systems can lead to queues of infinite length and to overcome this it is necessary to increase capacity so that it exceeds the mean demand made on it.

Exactly the same phenomenon would have happened in the HOTEL simulation, if the guests had queued; in fact there are similarities between the two simulations.

Structure of a Simulation

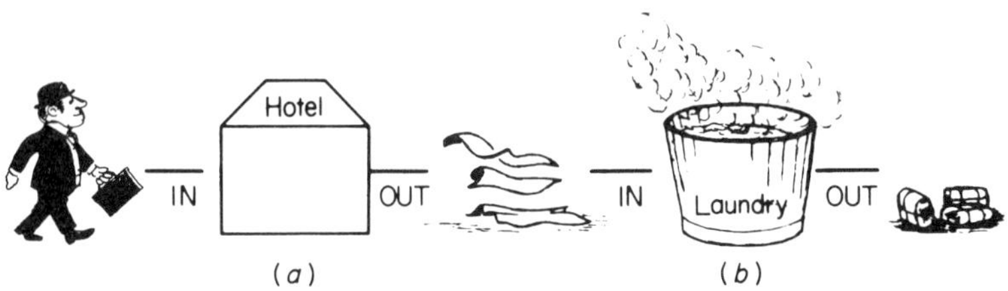

(1) Both deal with an INPUT stream:

(*a*) guests;
(*b*) dirty sheets.

(2) Both have RULES for dealing with these. These describe the model:

(*a*) accept no more than 4 guests each day. Excess guests must leave the system;
(*b*) accept no more than 4 sheets each day except Sunday. Excess sheets must queue.

(3) The EFFECT of the system RULES on the input stream is shown by the values of the ENDOGENOUS variables:

(*a*) number of guests staying,
number of guests turned away,
number of vacant rooms;
(*b*) number of sheets washed,
number of sheets not washed,
spare capacity of laundry,
number of sheets queueing.

The endogenous variables are most interesting because they show how well the system is working: they can highlight the faults and often provide clues to a remedy. In the case of the laundry it seems that long queues of dirty sheets are building up and the solution is to increase the capacity of the laundry. Before we accept this as our solution, however, we must be careful.

We would not normally base any statement on the simulation of a mere 20 days' guests. The particular 20 days' guests generated by the 'spinning-wheel' method may have been 'odd' for some reason. In fact, they were odd – they did not exactly follow the probability pattern and the mean was a bit low. The system once designed must handle the requirements of ALL guests EVERY day of the year. A much larger input stream representing thousands of days must be 'fed into' the model. When talking in terms of generating thousands of arrivals the spinning-wheel method loses some of its appeal. Fortunately, we can convert it to something less cumbersome.

Simulating using Random Numbers

Here is the probability pattern again:

x	0	1	2	3	4	5	6
$p(x)$	0·04	0·02	0·04	0·20	0·35	0·20	0·15

Here is the wheel:

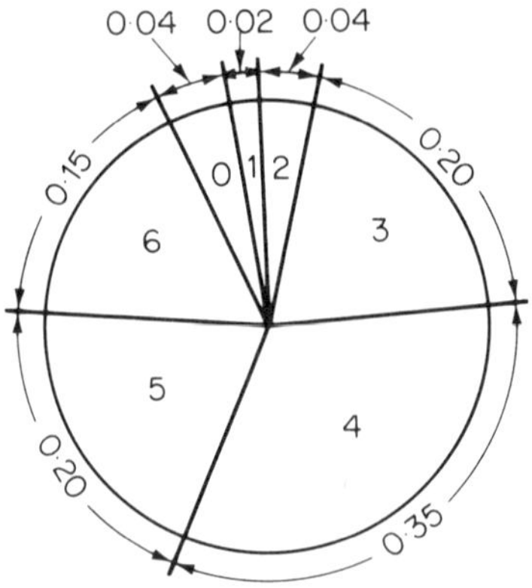

This time the wheel has some spokes and the lengths between each spoke are marked.

Now roll the wheel in a straight line. This is the impression it would make:

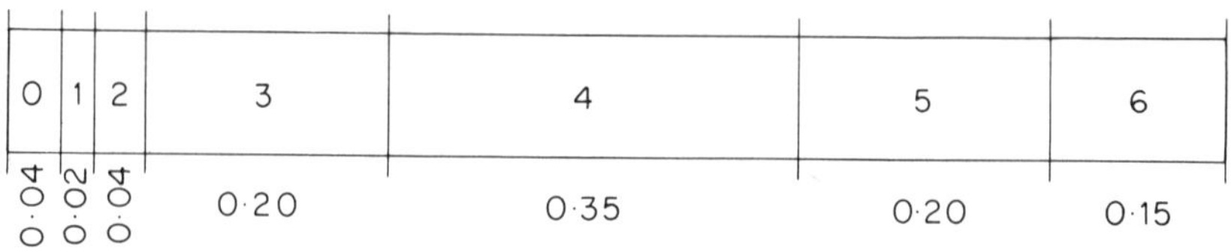

Rather than spinning the wheel, an alternative (but equally clumsy) way of generating the numbers to represent the number of guests arriving, would be to drop a pin from a height so that it fell on to the line: the number it falls on would be the number of guests. Now, if you add the distances underneath the line, you can see that the line has unit length. The diagram below shows the distances from A.

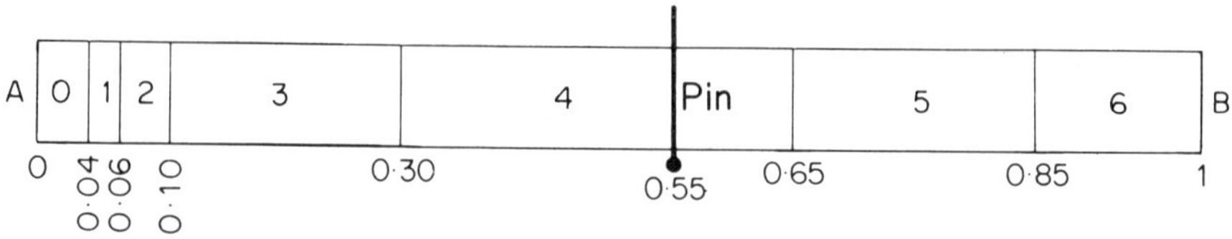

In the diagram, the dropped pin crosses the line at the distance from A measured on the axis AB; in this case 0·55. If the pin is dropped many times, every distance is equally likely to come up because the pin has no preference for its landing position. Every distance is between 0 and 1. If the distance is between 0·30 and 0·65 we record that 4 guests arrive. Each distance can be converted into a number of guests using the table over the page.

Conversion table

Distance	0–0·04	0·04–0·06	0·06–0·10	0·10–0·30	0·30–0·65	0·65–0·85	0·85–1
Number of guests	0	1	2	3	4	5	6

On this page are some RANDOM SAMPLING NUMBERS. If you put a decimal point in front, each number is between 0 and 1. The numbers have been generated in such a way that each one is equally likely to occur – not by dropping a pin or spinning a wheel but by an electronic method similar to this. You can find out more about this from K. Tocher's book, *The Art of Simulation.*

These numbers are readily available in books of statistical tables and can be used instead of the distances. It is as if someone has already spun the wheel or dropped the pin. Each random number can be converted into a number of guests using the conversion table. It is now possible to generate guests arriving for thousands of days with very little effort. It doesn't matter whether you take the random numbers from columns or rows provided that once you begin to read by columns say, you continue to do so. If the simulation is being done on a computer the table of random numbers could be stored for use

Random sampling numbers

34	34	32	22	60	53	91	17	33	26	44	70	93	14
10	55	35	25	24	28	20	22	35	66	66	34	26	35
97	26	33	94	42	23	01	28	59	58	92	69	03	66
88	08	19	85	08	12	47	65	65	63	56	07	97	85
43	39	76	93	08	79	22	18	54	55	93	75	97	26
75	73	00	11	27	07	05	20	30	85	22	21	04	67
17	27	31	42	64	71	46	22	32	75	19	32	20	99
70	40	46	55	46	12	24	32	36	74	69	20	72	10
85	33	75	18	88	71	23	44	54	28	00	48	96	23
00	79	91	22	29	01	41	39	51	40	36	65	26	11
68	36	24	72	03	85	49	24	05	69	64	86	08	19
32	59	51	82	86	43	73	84	45	60	89	57	06	87
05	88	78	44	63	13	58	25	37	11	18	47	75	62
90	80	77	80	26	89	87	44	23	74	66	20	20	19
23	20	33	62	62	19	29	03	94	15	56	37	14	09
54	55	38	57	94	62	68	40	26	04	24	25	03	61
78	41	60	58	10	60	88	46	30	21	45	98	70	96
44	13	45	05	34	59	75	85	48	97	27	19	17	85
60	77	90	91	60	90	79	62	57	66	72	28	08	70
02	07	19	40	21	29	39	45	90	42	58	84	85	43
72	40	73	05	50	90	02	04	98	24	05	30	27	25
78	30	79	47	96	92	45	58	40	37	89	76	84	41
01	88	69	57	54	45	69	88	23	21	05	69	93	44
33	79	96	84	28	34	19	35	28	73	39	59	56	34
78	39	73	88	62	03	36	00	25	96	86	76	67	90

although, because of storage limitations, we don't do so. Instead, we calculate each number from a formula, specially designed to yield numbers with the characteristics of random numbers. These are called PSEUDO-random numbers.

Whether the simulation is done manually or on a computer, the question arises: 'How many days should we simulate?' The answer is: 'Enough to represent the full variation of the system.' The statistician can provide the answer and the manager should always ask him, because simulations can be expensive and he is in charge of cost control.

Solution to the Linen Problem

If the results from the longer simulation show that the laundry IS a bottleneck then the capacity can be increased, not in reality but in the simulation. The input stream of guests arriving can be 'fed' into the changed model and we can see what the results will be. Turn back to the laundry simulation. The last three columns (the endogenous variables) will reflect the changes made to the system. It is useful to

Effect of introducing a Tuesday evening shift

	Day	*Arrival of sheets*	*Sheets queueing for wash*	*Sheets washed*	*Sheets not washed*	*Spare capacity number of sheets*
Tues	2	3	—	—	—	—
Wed	3	3	3	3	0	1
Thurs	4	4	3	3	0	1
Fri	5	3	4	4	0	0
Sat	6	3	3	3	0	1
Sun	7	0	3	0	3	0
Mon	8	7	3	3	0	1
Tues	9	4	7	7	0	1
Wed	10	4	4	4	0	0
Thurs	11	4	4	4	0	0
Fri	12	0	4	4	0	0
Sat	13	3	0	0	0	4
Sun	14	0	3	0	3	0
Mon	15	7	3	3	0	1
Tues	16	4	7	7	0	1
Wed	17	4	4	4	0	0
Thurs	18	4	4	4	0	0
Fri	19	4	4	4	0	0
Sat	20	0	4	4	0	0
Sun	21	0	4	0	4	0
Mon	22	3	4	4	0	0
Tues	23		3	3	0	8

consider how a system can be changed. The laundry capacity can be increased in three ways:

(1) Replacing all or part of the existing machinery by faster or larger machines.
(2) Buying extra machinery and hiring more staff.
(3) Increasing the machinery running time by introducing an extra shift in the evening.

The third way is probably the cheapest, but if the laundry is considering replacing its machinery in a year or two it might be appropriate to replace now. Let us consider the effect of an extra shift. The most obvious day for this is Tuesday, when Sunday AND Monday sheets are due to be washed. A shift which laundered 4 sheets would increase the average daily laundry capacity to 4, which is greater than the average arrival rate of dirty sheets (3·4 per day) for these 20 days. If you carry out the laundry simulation you will find that this extra capacity is sufficient to 'work off' the queue of sheets.

However, to determine the correct capacity increase it is necessary to simulate over a longer period in order to include the full variations of the input streams.

Once the bottleneck has been eliminated, the sheets are flowing smoothly round the system and the hotel can predict what clean sheets it will be receiving from the laundry. If the vans deliver every day except Sunday, then when the Tuesday evening shift is operating the hotel will receive the following number of sheets from the 3rd to the 23rd day:

Mon	*Tues*	*Wed*	*Thurs*	*Fri*	*Sat*	*Sun*
—	—	3	3	4	3	0
3	7	4	4	4	0	0
3	7	4	4	4	4	0
4	3	—	—	—	—	—

The average is of course 3·4, equal to the average number of dirty sheets. If the hotel were to rely solely on deliveries for its sheet supply it would surely run short. It is obvious that it needs a stock of at least 8 clean sheets to cover the Saturday and Sunday guests when no sheets are delivered. The total sheets required is greater than 8, of course, because during Saturday and Sunday there are some dirty ones at the laundry.

The position on Sunday (14th day) is this:

There are 3 guests arriving to stay requiring 3 sheets.
There is no delivery of clean sheets at the hotel.
Four sheets are dirty at the hotel from yesterday's (Saturday's) guests.
Three sheets are dirty at the laundry waiting to be washed.

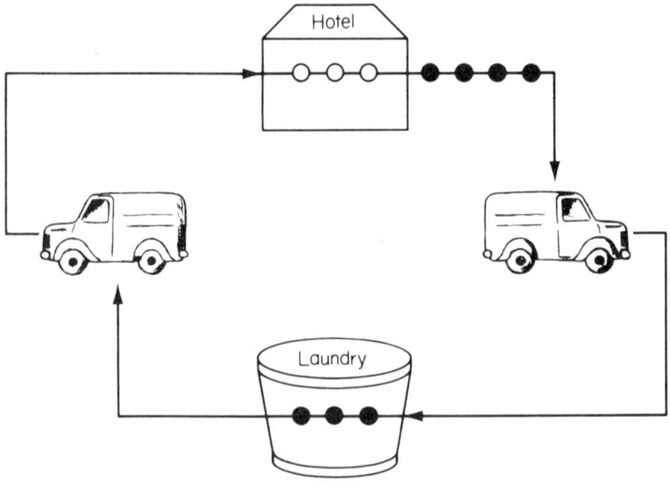

'The Sheet Situation' on Sunday, the 14th day

The total number of sheets in the system must be 10. The information on how many sheets are in each part of the system EACH DAY is contained in the simulation tables. For example, on Tuesday, the 16th day, 4 dirty sheets have arrived at the laundry, 7 have been washed and returned to the hotel, and 4 guests have arrived. At the end of the day the sheet situation is as follows:

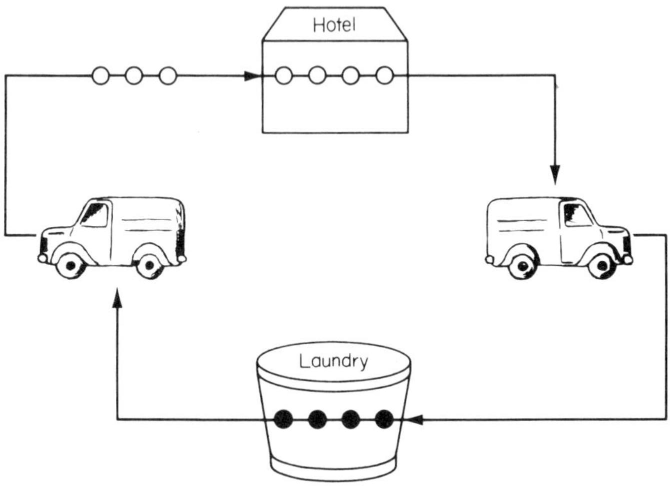

'The Sheet Situation' on Tuesday, the 16th day

This time 11 sheets are needed in the system. If this information is extracted from the tables for each day of the long simulation, a bar chart of sheet needs could be constructed.

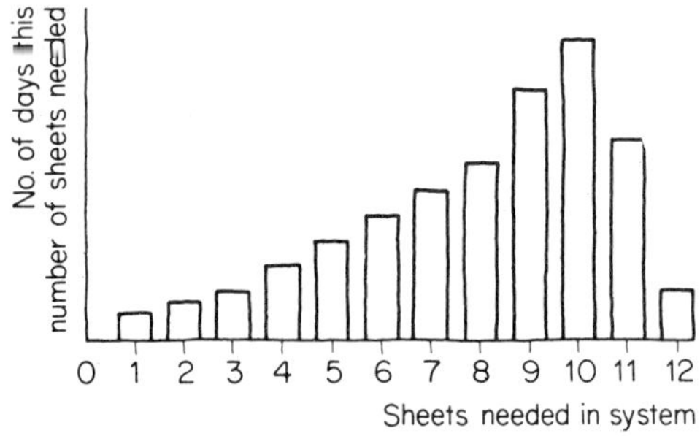

Sheets needed in system

With the laundry facilities represented in the simulation, the bar chart shows that 12 sheets are needed. Allowances must be made for machine and van breakdowns if these are considered significant.

We do this by 'inputting' a breakdown randomly at the rate equal to that observed in practice. Similarly, pilfered sheets can be 'stolen' from the system or worn-out sheets 'removed'.

A further piece of information that can be obtained from the tables is the TIME sheets take to get round the system. A batch of sheets can be 'flagged' and traced on their journey from hotel through collection, laundry and back. For example, the 3 sheets put on the beds for Thursday (the 4th day) arrivals reach the laundry on Friday, are washed and returned on Saturday, taking 2 days to complete the cycle. The sheets used on Friday (the 5th day), however, are not returned to the hotel until Monday, a cycle time of 3 days. A histogram of cycle time would look something like this:

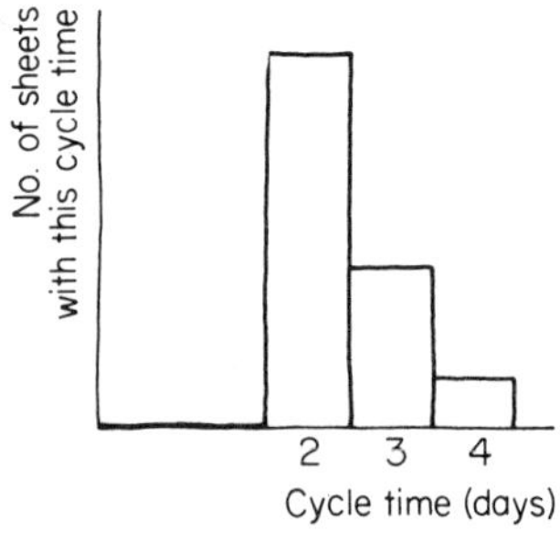

It is impossible with the current system to have a cycle time of less than 2 days. If the system were speeded up, however, by getting the sheets to the laundry so that they could be washed the SAME day, then the cycle-time histogram would look like this:

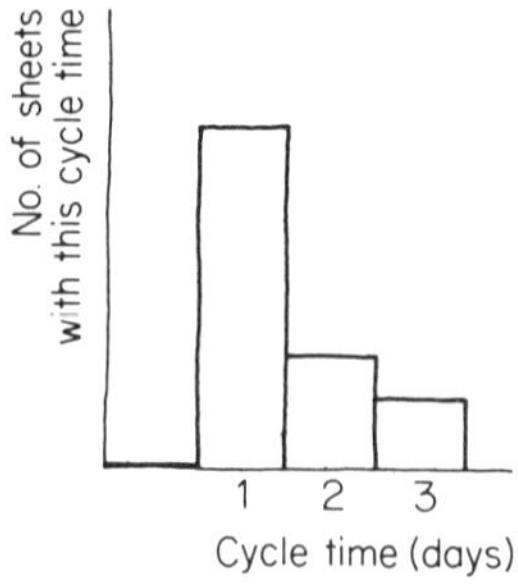

and FEWER sheets would be needed. Whether this is worth while depends on the cost of extra chambermaids needed to strip the beds in the shorter time, the cost of extra deliveries and the savings of having fewer sheets. It might be possible to arrange for *some* of the sheets to be washed the same day and this possibility can be investigated using the simulation method, by adjusting the model.

We have determined the number of sheets the hotel needs using the reduced and simplified model developed on page 109. The modifications necessary to treat a model of realistic behaviour are not difficult to understand but they complicate and extend the solution. We assumed that the arrival pattern of guests was the same whatever day of the year they arrived on, and that all stayed only 1 night. In fact, from Chapter 2, we know that different numbers of guests arrive on different days of the week (and probably different weeks of the year) and that their duration of stay varies. We would thus need to 'generate' two sorts of data for each day:

(i) number of guests arriving that day;
(ii) length of stay of each guest.

These could be generated from the appropriate probability patterns. The modifications necessary to increase the size of the hotel from 4 to something realistic is straightforward.

Modelling by Parts

Modelling is a very productive procedure. It forces us to examine the structure of a system and to break it down into manageable parts. Temporarily we concentrate on one part, for example the hotel, and forget about the others. It is then often possible to isolate those activities which relate only to this part and which are independent of the others. A selection of these activities can then be represented in a model component. However, always at the back of the mind are the connections with the other parts of the system and eventually these are considered, model components for them are developed, and all are linked together to assemble a complete system model. The problem cannot be *solved* by solving each model component independently. The *system* model is needed to represent the INTERACTIONS between the separate parts. Nonetheless, before any solutions are considered, modelling can highlight problems (for example, the laundry bottleneck in the linen system) and can provide clues to effective solutions (the Tuesday evening shift). The correct solution to a problem is not always the most obvious one: the answer to the problem of running out of sheets was not to buy more, because the laundry bottleneck would still exist, but to increase the laundry capacity instead. The model throws light on how a system works, displaying the relationship between the various features of a situation and, subject to the validity of the assumptions on which it rests, can demonstrate the effectiveness of a solution.

Models and their Solutions

A system model can take many forms depending on the complexity of the problem, how well it is defined and the variability of the situation. In Chapter 1, the well-defined problems of school were considered and a simple model with a simple solution was used to solve one. In Chapters 3 and 4, the equation $y = a + bx$, modelled the relationship between x and y, exhibited by the points on the graph.

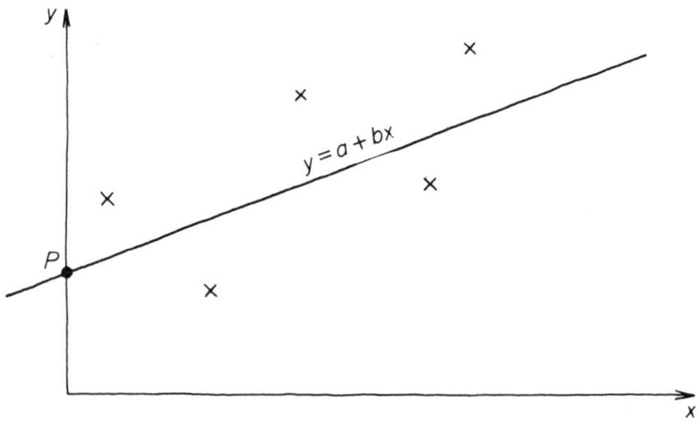

The behaviour represented by this data is one step away from the school-problem situation because it exhibits variation that cannot be fully explained by the line (or the equation). The solution to the problem (whatever the problem is) might be the value of y when x is zero, i.e. at *P*. In this case the answer is found by graphical analysis. Similarly, graphs were used to solve the model of rooms in Chapter 5. This model did more than just provide a solution: it indicated how the solution would change if the problem situation changed. The extent to which a model can respond to questions of the type 'what would happen if . . .?' is a guide to its usefulness. The existence of this facility when modelling (the changeable) real-life situations is as important as being able to obtain a solution in the first place.

Sometimes the model consists of a set of equations, as for the tin-can height problem of Chapter 4, and it is often possible to solve them directly, maybe using calculus. The system model on the first page of this chapter was solved directly in Chapter 5 to yield the prediction that sheets taken off the beds one day would be returned clean to the hotel the day after. Unfortunately, the behaviour is not as simple as the equations make out; they do not apply on some days of the week, e.g. Sunday. The assumptions necessary to apply these equations are not valid so the prediction is not valid. When a system exhibits variability so that it is difficult or impossible to write down an equation or set of equations that are applicable or when it is so complex that it is difficult to see or write down mathematically the relationship between the variables under your control and the variables you seek the values of, then the simulation approach, which involves a re-enactment of the system, may be helpful. It too has the 'what would happen if . . .?' facility. Possible changes can be tried out on paper (or on a computer) and assessed BEFORE putting them into practice. For example, in the simulation of the linen system, the effect of the Tuesday evening shift could be determined, before making the practical arrangements for, and incurring the cost of, an actual shift. It would also be possible to establish how many more rooms the hotel could have before the laundry would again become overburdened. The practical alternative of building more rooms, taking more guests and *observing* when the services become insufficient, is clearly absurd.

7
Coping with Uncertainty

Uncertainty of Statements

When people state an opinion, they tend to so do on the basis of their own, often fragmentary experience, sometimes augmented by the information contained in some recently collected data. Since the information on which most statements are based is at best incomplete, and at worst totally inaccurate, we must be uncertain of the truth of the majority of statements we encounter.

We must learn to recognise that ALL statements rely on data in one form or another, and it is very rare for a statement to be backed up by a perfect set of data – it would be impossible to collect. Our hotel manager has to make statements and collect data. Let us look at his problems.

He knows that a certain proportion of his guests who have made reservations will not turn up. He believes, and will STATE, that the proportion for this year will be about 10 per cent because it was 10 per cent last year.

Obviously he would like to know what proportion is not going to turn up today, but there is NO way of finding this out for CERTAIN. Even if he were to contact every person who had made a reservation, he could not rely on the information given because the people involved would not be able to foresee the unforseeable, such as an accident, which might prevent their arrival.

Clearly he could have acquired more information by collecting the data and this might have caused him to modify his figure of 10 per cent to, say, 12 per cent, but he still could not be CERTAIN that this figure was correct.

Statements, then, are imperfect approximations to the truth and are only PROBABLY correct. When using other people's statements for our decision making, it is essential to form a realistic assessment of the PROBABILITY that the statement is correct. Probability is a measure of the uncertainty associated with the statement and its study equips you to cope with uncertainty.

The problem the decision-maker faces is this: he must assign to any statement, whether he makes it or not, a single probability which expresses his degree of belief in that statement. This probability is the result of combining his previous convictions with the evidence supplied from the most recently collected data by the use of the laws of probability. To learn this skill we must study what probabilities are, how they are combined and how to interpret them.

Most people have an idea of probability: they use it in every-day conversation:

Number two: 'It'll rain tomorrow.'
Number one: 'Very likely.'
Number two: 'We will make over £2m next year.'
Number one: 'Probably.'
Number two: 'You will be Prime Minister before the year is out.'
Number one: 'Never!'
Number two: 'We will either make a profit, loss, break even or go bankrupt next year.'
Number one: 'Indupitably.'

The range of vocabulary describing chance seems to vary from never, through unlikely, possible, probable, very probable to certain. Now it is not easy to compare a 'very likely' with a 'highly probable', so to enable comparisons to be made, probabilities are measured on a scale 0 to 1 as we saw in Chapter 2. It is then easy to compare the chance of it raining tomorrow with the chance of a firm buying FORDS for its car fleet next year. Chance can be made a quantitative study.

A simple example should suggest certain general principles worth remembering when tackling uncertainty.

Suppose a committee is debating whether to hire temporary cover for an outdoor social event in October. Money is short and they only wish to go to the expense of hiring canvas if it is reasonably certain that people will need shelter from the elements. Past weather records indicate that the chance of rain on the chosen date is about $\frac{1}{3}$ and the chance of high winds is $\frac{3}{4}$. Somehow these two probabilities must be combined to give the committee the chance of the weather being inclement. The problem is to find the chance of the weather being windy or wet.

Obviously this chance must be greater than either $\frac{1}{3}$ or $\frac{3}{4}$ so multiplication and subtraction are ruled out as methods of combining them into a single probability. Try it and see. Since we are tackling the problem in an experimental frame of mind let us investigate addition:

$$\frac{1}{3} + \frac{3}{4} = \frac{4}{12} + \frac{9}{12} = \frac{13}{12}$$

Now 13/12 is 1, plus a bit. We have combined two probabilities (numbers less than 1) and produced a number larger than 1. This number cannot represent a probability so the method of combining the two probabilities must be wrong, even though adding did seem logical. What went wrong?

Let us look at the basic assumptions. What did that probability of $\frac{1}{3}$ really refer to? Days when it just rained, or did it include also days when it rained but there was some wind with it?

The $\frac{1}{3}$ must have referred to the probability of days when rain fell, including the days when there was wind with the rain. Similarly, the $\frac{3}{4}$ must be associated with days when the wind blew irrespective of whether it rained as well. BOTH probabilities, $\frac{1}{3}$ and $\frac{3}{4}$, included the probability of the same event, a wet AND windy day. There are FOUR distinct weather types for that day in October and they are set out in the diagrams.

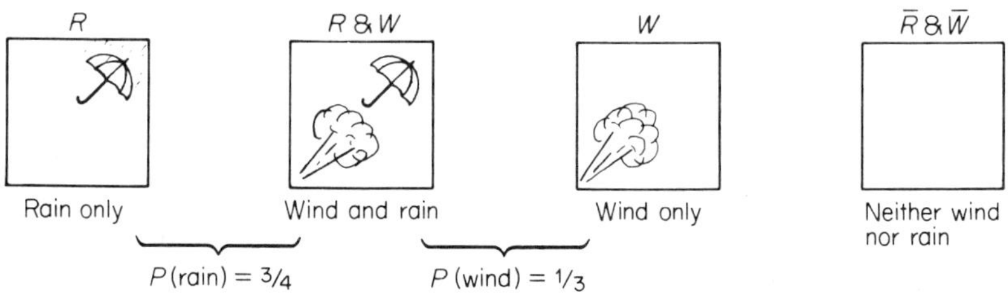

When we talk of rainy day we don't exclude the days when it was also sunny or also windy. Windy days are taken to mean days when there was wind, and days are not excluded just because rain fell while the wind blew. By adding the two probabilities, the probability of days which were both windy AND wet was added twice: once with the windy-day probability and once with the wet-day probability.

The correct method of combining the probabilities is

$$\text{Probability (rain OR wind)} = \frac{1}{3} + \frac{1}{4} - \text{Probability (rain and wind)}$$

This is a particular example of a general result written in statistics books as

$$P(A \text{ or } B) = P(A) = P(B) - P(A \,\&\, B)$$

There was not enough information given to solve the problem. We must request further information about the probability of days when it was rainy AND windy.

The general principle is: 'When defining an event, be sure that you know ALL the situations covered by that event.'

A further principle has been staring us in the face. Certainty has probability ONE! If the probability that it will rain is $\frac{3}{4}$, then it is CERTAIN that it either rains or it doesn't on a particular day, so

$$\text{Probability of rain} + \text{probability of no rain} = 1$$

or

$$P(\text{rain}) + P(\text{no rain}) = 1$$

which becomes

$$\tfrac{3}{4} + P(\text{no rain}) = 1$$

$$P(\text{no rain}) = 1 - \tfrac{3}{4} = \tfrac{1}{4}$$

Certainty can be a useful idea in answering probability questions. From our previous diagrams we see that there are FOUR states of the weather that cover ALL the possibilities:

R It rains but there is no wind
R & W It rains and it is windy
W There is wind but no rain
R & W There is neither wind nor rain.

ANY weather condition can be placed into one of the four exclusive states, R, R & W, W or $\bar{R}$ & $\overline{W}$, and NO weather condition falls into more than one state. Moreover, since one of these states MUST happen,

$$P(R) + P(R \,\&\, W) + P(W) + P(\bar{R} \,\&\, \overline{W}) = 1$$

We have decided on the values to be given to P(rain) and P(wind) so now let us assign a probability to the event (R & W) of $\frac{1}{4}$ and calculate $P(\bar{R} \,\&\, \overline{W})$.

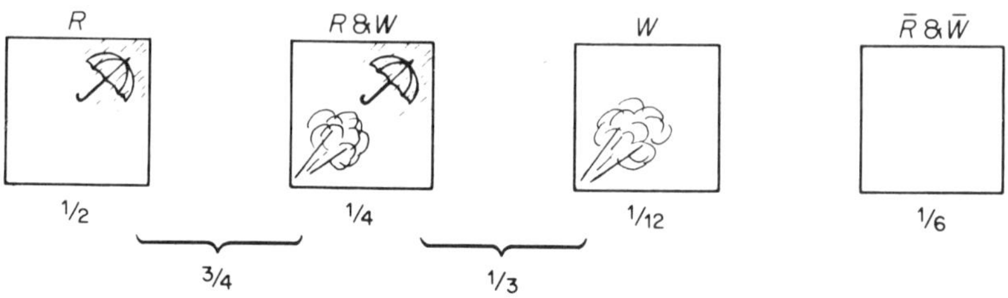

Set out the calculation for yourself using the formula for certainty.

One last question: A philosopher, down a sewer, sees the water level rise about his boots and conjectures, rightly, that this is due to rain. Using the knowledge we have about the past weather conditions, he sets out to work out the probability that it is windy too. How can he use his knowledge that it is raining already to modify his probability of it being windy?

First of all he reduces the number of states in the certainty equation to the first two, R and (R & W), since it IS raining:

$$P(R/\text{it is raining}) + P(R \text{ and } W/\text{it is raining}) = 1$$

It helps if you read the slash as 'given that'.

In words we have: 'The probability that it is raining with no wind, given that it is raining, plus . . . is one.'

Originally R was twice as likely as (R & W) so we can deduce that

$$P(R \text{ and } W/\text{it is raining}) = \tfrac{1}{3}$$

or, as it is more often written in statistics books,

$$P(\text{windy/rain}) = \frac{P(\text{wind \& rain})}{P(\text{rain})} = \frac{\frac{1}{4}}{\frac{3}{4}} = \tfrac{1}{3}$$

This formula is the basis for combining probabilities and leads directly to a very useful result due to the Reverend T. Bayes. You should read more about it starting with Dennis Lindley's book, *Making Decisions*, published by Wiley, 1970. To help you relate the ideas developed so far to those you will find in the other statistics books you will need to read, you might care to note the following.

Our EVENTS were R, W, R & W, $\bar{R}$ & $\overline{W}$, rain, wind.

Some events were MUTUALLY EXCLUSIVE.

Rain and W were mutually exclusive but R and rain were not. Rain was the event of R and (R & W) combined so

$$P(\text{rain}) = P(R) + P(R \,\&\, W)$$

Consult the diagrams to identify the events.

If two events, A and B, are INDEPENDENT then information about event B does not affect the way you think about the event A and we write:

$$P(A/B) = P(A) \qquad \text{(Read the slash as 'given that')}$$

but $$P(A/B) = \frac{P(A \text{ and } B)}{P(B)}$$

and so $P(A \text{ and } B) = P(A) \,.\, P(B)$ in the very special case when A and B are INDEPENDENT.

Usually $P(A \text{ and } B) = P(B)P(A/B) = P(A)P(B/A)$.

The rules for combining probabilities always hold. Everyone agrees with them and uses them. Interpreting the resultant probability is a different matter. We should recognise that the word probability means different things to different people.

To a casino manager, the probability of the casino winning in a game of roulette is 2·8 per cent, which means that over a long period of time he will take about 3 per cent of ALL the money put on the table. He isn't worried when one night the bank is almost broken, because he 'knows' that eventually he will recoup his losses because his wheel is SET to give him a 3 per cent profit. Here, then, we have a man who is happy with his probabilities so long as he has a house limit to ensure that in the random ups and downs of fortune an occasional large win for the opposition does not bankrupt him. His probability mechanism, the beautifully balanced roulette wheel, is stable, and stability of the probability mechanism is highly desirable when money is at stake. The market research man is not so fortunate: his probability mechanism is likely to be very unstable. What, he wants to know, is the chance that a person who said he would buy a product actually buys it? He could collect information about people and find that in the past the proportion of people buying, after they say they will buy, is 0·70, but what guarantees has he that this will remain constant for all products and all age groups?

The oilman needs to know the chance of a trial well being dry. He ensures that this is as small as possible by employing experts to advise him on geological structures, but in the long run he can only drill so many dry wells with the money he has available before he either goes out of business or has to put up his prices. How many dry wells are likely to be sunk before one is successful? Maybe he will have to suspend exploratory wells until the geologist's improved methods ensure that the probability of a dry well is small enough for the oilman to take the financial risk of drilling.

Our hotel manager looks at probability in terms of bookings. Hotels would never be full unless they overbooked because people cancel at short notice or just do not arrive. The manager will be trying to fill his hotel as often as possible so he adopts the policy of accepting more bookings than he has rooms, maybe up to 20 per cent more if he has the chance. Naturally there is a risk that not enough people will cancel and the receptionist will be faced with more guests demanding their rights to a room than there are rooms available, and the manager has to assess this risk and balance the money he makes by overbooking against the money or money equivalent necessary to placate a disappointed guest.

What probabilities has he to work with?

Does he count the total number of bookings made in a year, determine how many are actually taken up, t, and say that the probability of a booking being taken up is t/b? Or does he adopt a more sophisticated approach? Does he divide his customers into Foreign and British and calculate a separate t/b for each? Will he make separate calculations for different months? Or will he just rub his chin and think of a figure between 0 and 1?

The consultant statistician sometimes looks at probability ideas from a different point of view. When setting up a piece of consultancy the statistician often asks the impossible sounding question: 'How sure do you want to be that the result I give you will meet your specification?'

The answer is usually: 'Absolutely sure of course! What do you think you are paid for?'

The statistician then has to restrain himself from telling the client that absolute certainty (Probability = 1) would require the taking of a sample which includes everything, which would cost more than the client could pay.

If a SAMPLE of the data is taken the conclusion can only be that the answer obtained is *probably* correct. A sample of 3 would probably not provide such a good answer as that from a sample of size 300!

If the client was an accountant with the problem of determining the assets of an organisation, he would probably settle for a result from the statistician which was accurate to within 1 per cent of the true assets together with the statistician's assertion that he was 0·95 confident that the result WAS within 1 per cent of the true assets. The statistician cannot GUARANTEE his result is correct to within 1 per cent. All he is saying is, 'I am a decision maker, and as a decision maker sometimes I am right, sometimes I am wrong. If you pay me enough to collect a large amount of relevant data you can have a decision from my drawer which contains 99 correct decisions and one incorrect decision. The decision you get is either correct or not correct. But you come to me because you know that of, say, 200 decisions I have made using the 0·99 drawer, only about 2 will have been wrong and this is a chance you are prepared to take.' Fortunately, in this situation, even when the DECISION is wrong, the answer will not be too far out.

All these people view probabilities from a slightly different point of view: some like the casino manager are only interested in AVERAGE behaviour while the punter and the oilman are very concerned about the outcome of the next uncertain event. But no matter which group they belong to they COMBINE their probabilities by applying the same rules, and if when the rules are applied to the probabilities the result is absurd (a probability, say, outside the range 0 to 1) then it is immediately obvious that one or more of the original probabilities was not accurate for some reason. This is a useful method of testing whether people are consistent in assigning probabilities.

Now that you have a feel for some of the ideas of probability we can explore some of the applications.

Uncertainty when Sampling

One of the simplest but most powerful applications is in the control of quality. Unless products are carefully monitored during production, imperfect units are incorporated into larger units which are then rejected. Large units carry a large investment in time and material and if they are defective, scrapping them is expensive. Even if the defective part is located, this costs money and the large unit itself represents idle capital. Quality control is essential in every industry. Chapter 11 in the Pelican book, *Facts from Figures*, gives a good introductory account. The basis of statistical quality control is the assumption that the design specification of a component is being met on average, and the actual dimension varies within acceptable limits.

Let us take a simple example – the design and manufacture of a shaft for a pump. The designer of the pump decides that the shaft shall be a cylinder with the diameter being the critical dimension.

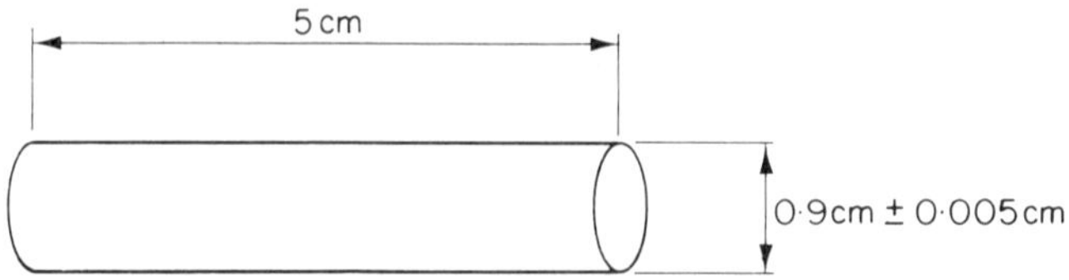

Designer's drawing

The lathe operator must work on a standard 1-cm bar and reduce it to 0·9 cm. It is impossible to produce a shaft with a diameter of 0·9000000000000 cm time after time on a lathe. A typical set of diameters might be 0·901, 0·902, 0·898, 0·899, which are well within the designer's tolerance limits of 0·895 to 0·905 cm. Usually the operator meets the tolerance limit requirement with quite a bit to spare.

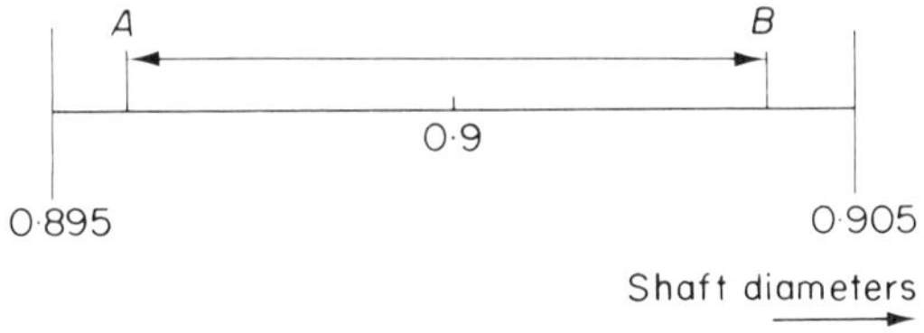

AB is the range of diameters produced by the lathe operator when he is 'on form'. He will be aiming to produce his shafts with a diameter of 0·9 cm so most of them should have diameters close to 0·9 cm and only a few should have diameters as low as *A* or as high as *B*. If we plotted a point for each diameter we would eventually construct a familiar pattern.

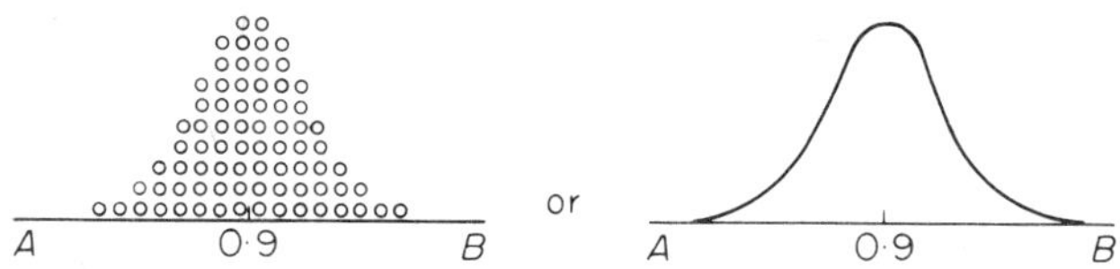

All this assumes that the operator was on form, the lathe was set to produce shafts with diameters of 0·9 cm and these diameters were produced at random, NOT in the sequence

0·897, 0·897, 0·898, 0·898, 0·898, 0·899, 0·899, 0·899,
0·899, 0·900, 0·900, . . ., 0·902, 0·902, 0·903

which would indicate that successive shafts had larger diameters. These results can be plotted with time as the horizontal axis.

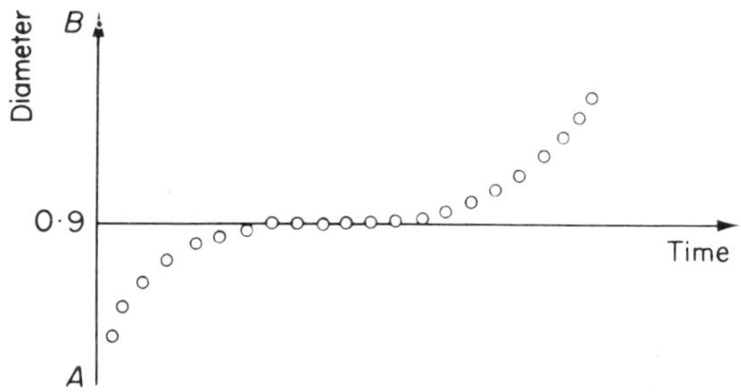

Such a plot representing an increasing diameter would indicate that our degree of belief in the idea that the shaft diameter was still conforming to a mean of 0·9 cm should be modified in the light of the data in favour of the alternative idea that the diameter was greater than 0·9 cm on average. When enough data indicating a change has been collected, the lathe would have to be adjusted.

Plotting data can be very instructive. Keeping a situation 'in control' is a very necessary preoccupation of most managers. They need to know when a seemingly large value is a result of a change in the situation or just a manifestation of the variation inherent in the situation. To exercise their control they need to see individual values in the context of past data and the expected variation.

Our hotel manager, for instance, would keep a chart of his food bill to keep track of rising prices.

He starts by talking to himself: 'Experience leads me to believe that my food bill is on average £500/week for the summer but I would expect it occasionally to fall to £400 or exceptionally rise to £600 even when there was no special catering.'

Then he draws a chart. Many people reading statistics books read about drawing charts, answer questions about drawing charts, say how essential and illuminating chart-drawing is BUT never actually draw one. Draw this one:

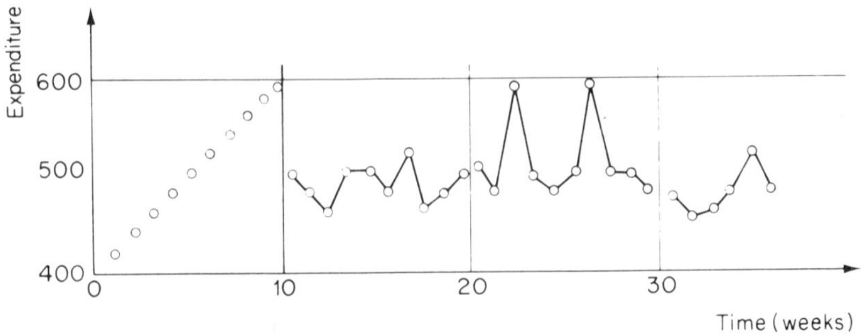

In the region 0–10 the situation is going out of control and action, such as sacking the catering manager or changing the supplier, is called for. On the evidence of the data recorded in the region 10–20 the manager has no reason to modify his idea that the mean cost is £500. In the third period the situation has changed again: there are two weeks with very high costs but they are still within the limits laid down. The cause for concern lies in the probability of getting two such high values in seven values. Back to our model of the variation.

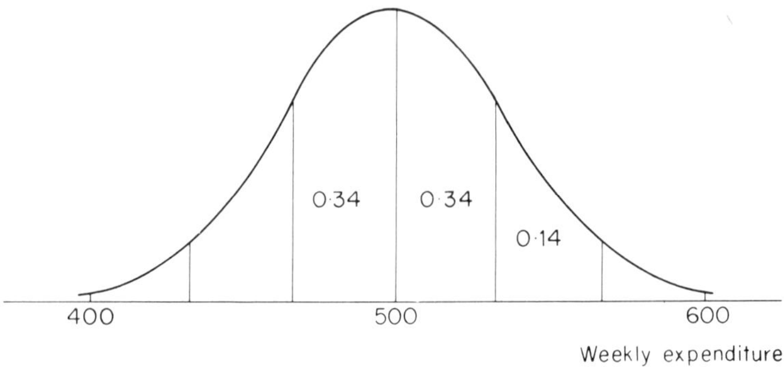

Divide the range 400–600 into 6 equal parts. The chance of getting ONE value in the highest section, 566–600, is about 1/40. Work it out by eye. How many times will the small area in the model fit into the overall shape?

The probability 1/40 is another way of saying in 40 weeks we would expect about 1 value as high or higher than 566. Consequently if there were no values higher than 566 in a 40-week period we would not be surprised. Even 2 in 40 weeks would not cause comment but 2 in 7 weeks is remarkable, IF the situation giving rise to a mean of £500 has not changed. It could happen, of course, that you win a premium-bond prize twice in 7 weeks from your solitary bond, but it is extremely unlikely; rather like getting a hole in one at golf, twice in a lifetime!

What, then, should our manager do? He suspects, from his data, that the situation is different now from what he expected. If he is of a nervous disposition he will rush around looking for a cause of the change but if he is more phlegmatic he will wait another week or two to check that the first set of results was reliable. Notice that both types have noticed an unexpected result but their managerial responses are quite different. Decide what type of manager you would make by analysing your attitude to improbable results arising in 'significance testing' (look up significance testing in an introductory text like *Statistics made Simple*).

Would you ever ACCEPT the truth of a NULL HYPOTHESIS because of insufficient evidence? Would you automatically reject a null hypothesis if you obtained an unlikely result?

What would you do if a value was not quite unlikely? If you cannot decide, discuss the question with your lecturer or statistical adviser.

Data in statistical thinking modifies the degree of belief we have in a certain proposition. If, as a market researcher, you believe that your firm's product is bought by 20 per cent of the population, and a properly conducted survey indicates that only 10 out of 400 actually buy the product, then your belief in the figure of 20 per cent must be severely shaken, and your degree of belief considerably reduced. At the same time your degree of belief in a lower percentage will increase. In the quality-control situation, action must be taken on the basis of degree of belief, which is constantly being modified by the data collected. To examine degrees of belief in practice, consider Fred, a custodian of the incoming quality of semi-conductors which arrive at the factory in batches of 1000.

A Model for Uncertainty

Fred's firm has entered into an agreement with the suppliers. The suppliers say that less than 5 per cent of their semi-conductors are defective when they leave the factory. Fred's firm trusts the supplier implicitly. (Degree of belief in the supplier's assertion that less than 5 per cent are defective = 1.) However, the firm reasons that, even with the best of intentions at the supplier's end, a 'bad' batch will be despatched occasionally. Consequently the firm proposes that a certain number, n semis, from each batch be inspected on arrival.

The data arising from this inspection will be used to modify the firm's degree of belief that the supplier is maintaining his standards. Let us be specific. Suppose that each batch contains 1000 semis. On average there should be less than 50 defectives in the batch IF we believe the supplier. How is the firm to check quickly the quality of the batch? It could test electronically all 1000 semis, but this would be expensive; or it could take a single sample which would be cheaper. But you never get anything for nothing. The saving of the test costs in taking a sample is paid for by the degree of uncertainty about the quality of the batch based on the information in the sample.

What action should be taken if the randomly selected semi is a dud? Decide that the batch is a 'bad' one and return it? You might well be returning a batch which had only ONE defective in it: the one you selected. The suppliers are not going to like that. A satisfactory semi would lead to the firm accepting the batch but that too is risky – for the firm this time. The one selected could, just possibly, be the only 'good' semi in the batch.

Making a decision based on a sample of size one is obviously risky, but cheap. On the other hand inspecting every semi is very expensive but there is less chance of a mistake. There must be a balance between the cost of testing, the cost of returning 'good' batches and the cost of accepting 'bad' batches. Let us revise the situation.

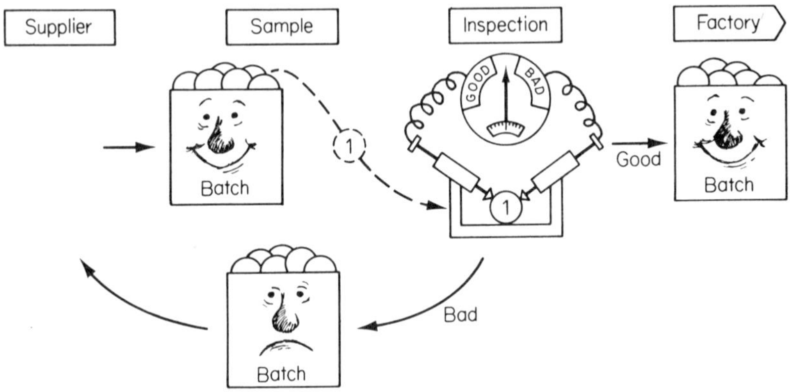

The batches are either as good as the supplier says or they are not. Whether they are or not, the firm will return some good batches because they are thought to be bad, and accept some bad in the mistaken belief that they were good. Both supplier and firm need to know how often a wrong decision is made.

Hopefully, the decisions should follow the pattern of the table below:

Decision	*Batch* — *Good*	*Batch* — *Bad*
Accept	Often	Rare
Reject	Rare	Often

Having made the very large step of admitting that a decision could be wrong, we must now decide how to act in a state of uncertainty. Our methods must be organised to ensure that the OFTEN in the table is very often, and the RARE as rare as we can afford to make it.

Another way of expressing this is to draw a graph demonstrating how we would choose our sampling method to decide how good the batches are.

In the earlier chapters we have discussed the essential features of a graph. The variables first of all: there must be at least two and the independent one, which will be represented in the horizontal axis, will be the variable over which the firm has no control – the TRUE proportion of defectives in the batches. Probability will be the dependent variable: the probability that a batch will be passed as a result of the firm's inspection process. You will, of course, now be waiting to be told the scale of both axes.

The *proportion* of defectives will obviously range from a fairly low value to a higher one and probability has a fixed range of values anyway. We leave the details to you to fill in on your sketch.

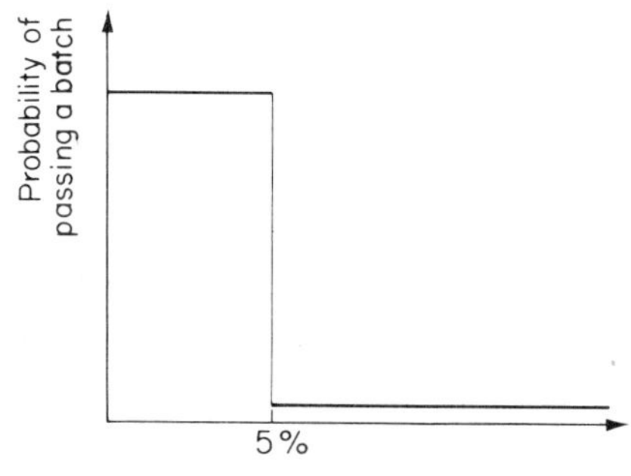

Fill in the scales before you read on!

When the percentage of defectives is very low we would like the sampling scheme to pass ALL the batches. The probability of a 'good' batch passing when the percentage of defectives is low should be as high as a probability can go. BUT, as soon as the supplier slips in a batch with a high proportion of poor quality material the inspection system should NOT pass this batch. The probability of passing such a batch should be as low a probability as it is possible to achieve. Summarised statistically we would write

$$P(\text{pass}/p \text{ is 'good'}) = 1$$
$$P(\text{pass}/p \text{ is 'bad'}) = 0$$

This graph is an ideal to strive for but unfortunately not realisable in practice as we shall see.

By adopting suitable inspection strategies we might get as close to the ideal as the diagram below suggests.

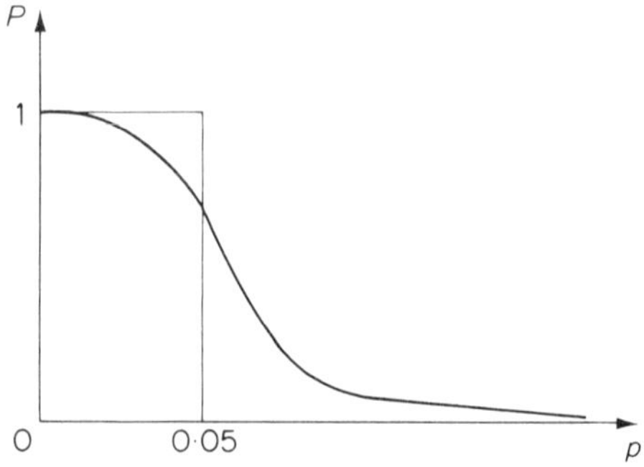

Analysing the graph piece by piece we see that for low values of p (good values from the firm's point of view) the probability of passing the batch is high, but as p increases the lack of belief in the supplier's claim is reflected in the steep drop in the probability of acceptance after the critical value of $p = 0{\cdot}05$ is reached. When p is very much bigger than 0·05 the chance of acceptance is negligible.

What are the implications so far of our non-mathematical study of the system? Clearly both firm and supplier must draw up an agreement about the inspection method used, because the supplier will occasionally receive back perfectly acceptable batches. The supplier will not wish to go to the expense of inspecting EVERY returned semi to decide whether the batch WAS of acceptable quality. Much better, he accepts the situation philosophically and sells the returned batches at a cheaper price admitting that they MIGHT contain a fair number of defectives.

Make a list for yourself of the difficulties the firm's inspection system faces. Put yourself in the position of an inspector of the quality of iron ore arriving in railway wagons, or a maintenance supervisor receiving batches of machine parts needed to keep a continuous process working, or a quality controller for rolls of aluminium foil for food wrapping.

Ask yourself:

What is the object of the inspection procedure?
How big a sample should be taken?
How do I sample effectively the material to be tested?
What do I do with a defective when I find one?

Now let us investigate quantitatively the consequences of a very simple inspection scheme. We start by finding just one point on the graph of probability of acceptance against quality.

Initially, we will consider a situation where the supplier batches his material, 5 units to a batch, and agrees with the firm that $p = \frac{1}{2}$ represents acceptable quality. The firm adopts the inspection strategy of selecting a unit at random from a batch, testing it and accepting the whole batch if that unit was not defective. What happens when p IS $= \frac{1}{2}$?

One half of all the semis sent out by the supplier will be defective. They are batched randomly in fives and some of these batches will have 0 defectives, others just one, or two, or three or four or, very rarely, FIVE. The number of defectives varies from 0 to 5, rather like the number of days guests stayed in the hotel varied. But this time, instead of observing guests, collecting data about their length of stay and drawing a histogram, we will use a PROBABILITY model to draw the histogram.

Building the model

Method

After the first semi had been packed, it was either good or bad; if the first was good, the second could be good or bad and so on. Careful drawing indicates quite clearly that there are 32 routes through the diagram opposite.

There is only one route which gives the result of 5 bad semis, so the probability of this is 1/32 because all routes have the SAME probability of occurring when $p = \frac{1}{2}$.

Carefully draw the complete diagram – you might need more than one attempt to get a clear picture – and trace each of the 32 routes,

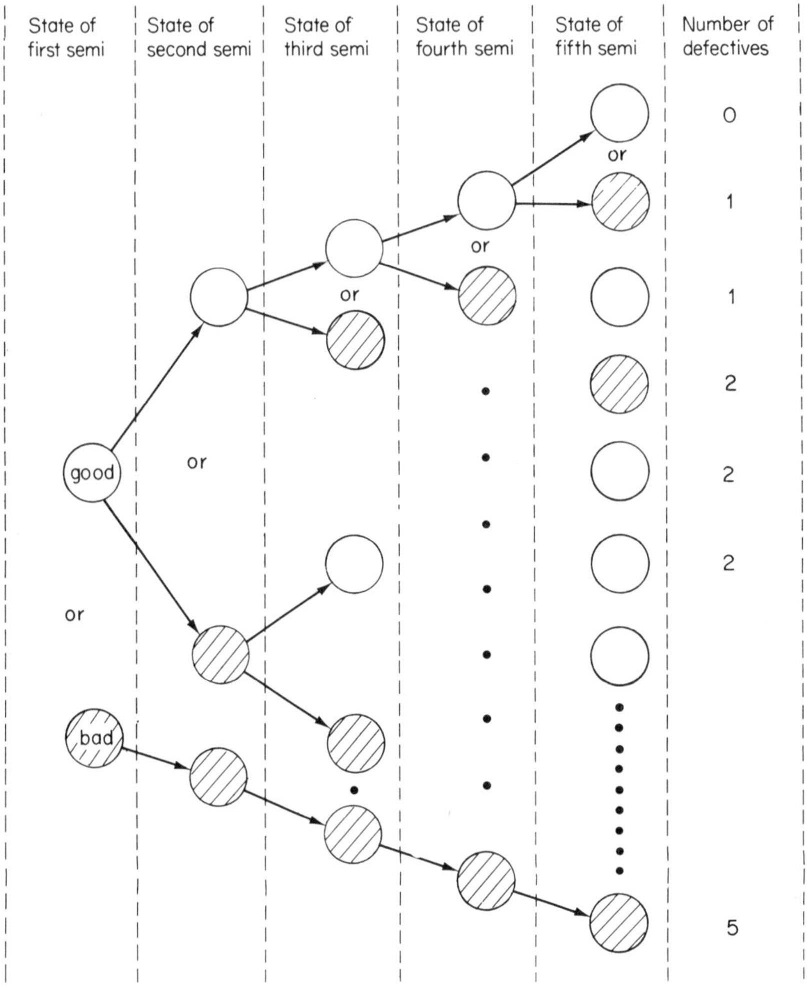
State of first semi
State of second semi
State of third semi
State of fourth semi
State of fifth semi
Number of defectives
good
bad
or
0
1
1
2
2
2
5

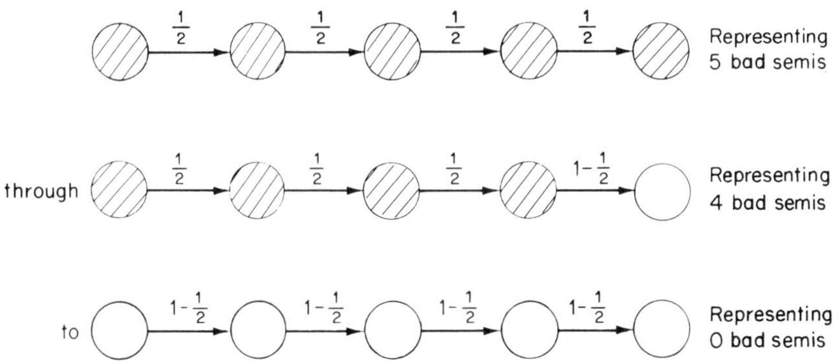
$\frac{1}{2}$
Representing 5 bad semis
through
$1-\frac{1}{2}$
Representing 4 bad semis
to
Representing 0 bad semis

and record separately the number of defectives encountered on each. These will vary from 0–5 depending on the route. Arrange these numbers into a histogram and check that it corresponds with the one given in the text.

Another way to determine the probability of getting a particular number of defective articles in a batch is to look at the way the packer fills his batch. He fills each batch with 5 articles, one at a time. He is not aware which articles are electronically defective and which are not; as far as he is concerned he is taking part in a lucky dip, selecting articles in turn from a limitless supply, half of which are defective. We can follow his selection procedure by tracing routes through the diagram. Consider the single route which leads to a batch with 5 defectives.

First of all the packer selects a defective to fill the first position in the batch: the chance of selecting this is $\frac{1}{2}$. He selects from the supply of articles again and gets another defective. This choice was completely INDEPENDENT of his previous choice, and the probability that the second article to be packed is defective is $\frac{1}{2}$. Since the 5 separate events,

the first article selected was defective
the second article selected was defective
. . .
the fifth article selected was defective

are all INDEPENDENT, the chance of 5 defectives,

$$P(5) = \tfrac{1}{2} \cdot \tfrac{1}{2} \cdot \tfrac{1}{2} \cdot \tfrac{1}{2} \cdot \tfrac{1}{2} = \tfrac{1}{32}$$

exactly the same result as before.

Each of the 32 different routes through the diagram can be assigned a probability. For instance, the route

which is part of the event '4 defectives', has probability $(1-\frac{1}{2}) \cdot \frac{1}{2} \cdot \frac{1}{2} \cdot \frac{1}{2} = \frac{1}{32}$. There are other routes which are part of the event '4 defectives in a batch'. Trace them on the diagram. Each of these 5 routes has probability $(1-\frac{1}{2})^1 (\frac{1}{2})^4$ so we can write

$$\begin{aligned} P(4 \text{ defectives}) &= (1-\tfrac{1}{2})^1 (\tfrac{1}{2})^4 + \tfrac{1}{2}(1-\tfrac{1}{2})(\tfrac{1}{2})^3 + \ldots + \tfrac{1}{2}^4(1-\tfrac{1}{2})^1 \\ &= 5(1-\tfrac{1}{2})(\tfrac{1}{2})^4 \\ &= \tfrac{5}{32} \end{aligned}$$

The event '4 defectives' was made up of the 5 MUTUALLY EXCLUSIVE events;

First article non-defective AND the rest defective
Second article non-defective AND the rest defective
. . .
Fifth article non-defective AND the rest defective

Calculate $P(0)$, $P(1)$, $P(2)$, and $P(3)$ using the same reasoning. Draw the routes each time and write down the argument in full. You will use this important method of reasoning later to calculate the probability that the batch will be accepted when p is different from the value $p = \frac{1}{2}$.

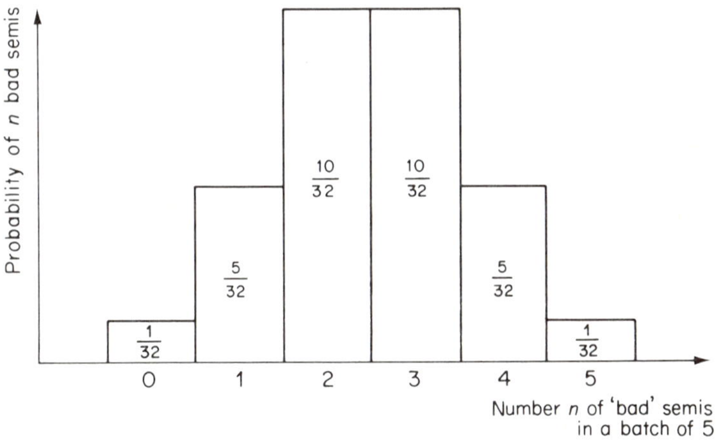

Histogram of the numbers of 'bad' semi-conductors in a batch of 5 semi-conductors

Reading this histogram carefully we see that only $\frac{1}{32}$ of all the batches have 0 defectives while $\frac{5}{8}$ of them have 2 or 3 defectives in them.

Take a batch with 3 defectives. What is its chance of getting through inspection? Well there are 5 semis, 2 are not defective and 1 is to be selected at random. The probability of a non-defective is $\frac{2}{5}$. So $\frac{2}{5}$ of all the 3 defective batches will get through. Altogether, the proportion of batches which get through when $p = \frac{1}{2}$ is:

$$P(A/p = \tfrac{1}{2}) = P(0 \text{ defectives}) \times P(\text{selecting a good semi}/0 \text{ defectives})$$
$$+ P(1 \text{ defective}) \times P(\text{selecting a good semi}/1 \text{ defective})$$
$$+ P(2 \text{ defectives}) \times P(\text{selecting a good semi}/2 \text{ defectives})$$
$$+ \ldots$$

$$P[A/p = \tfrac{1}{2}] = \tfrac{1}{32} \times 1 + \tfrac{5}{32} \times \tfrac{4}{5} + \tfrac{10}{32} \times \tfrac{3}{5} + \tfrac{10}{32} \times \tfrac{2}{5} + \tfrac{5}{32}\tfrac{1}{5} + \tfrac{1}{32} \times 0$$
$$= 5 \times \tfrac{1}{32}[5 + 20 + 30 + 20 + 5] = \tfrac{16}{32} = \tfrac{1}{2}$$

Exactly half the batches get through when $p = \frac{1}{2}$.

The supplier would be livid! More than $\frac{1}{2}$ of his batches which were of acceptable quality are being returned.

By changing the value of p and calculating the probability of acceptance, $P(A/p)$, the qualities of the sampling scheme can be assessed. Ideally, you remember, the curve should be Z shaped, but in this case the plot is a straight line.

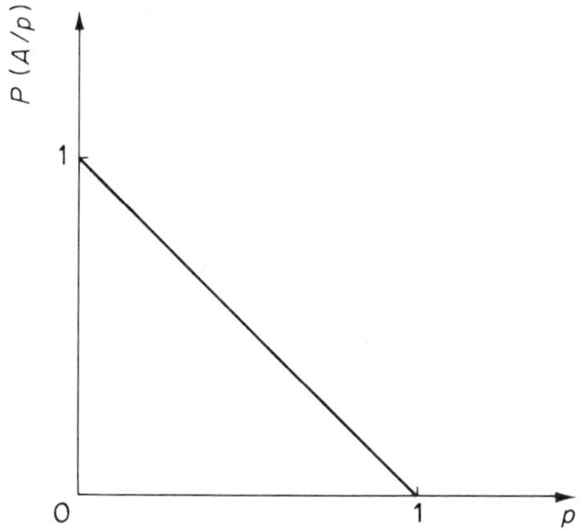

If you recall the straight-line theory from Chapter 3, you will deduce from the diagram that the equation of the line is $P(A) = 1 - p$.

By substituting p for $\frac{1}{2}$ in the expression for $P(A/p = \frac{1}{2})$ we have

$$P(A/p) = (1-p)^5 + 5p(1-p)^4\,0{\cdot}8 + 10p^2(1-p)^3\,0{\cdot}6 + 10p^3(1-p)^2\,0{\cdot}4 + 5p^4(1-p)0{\cdot}2$$

which DOES simplify to $P(A) = 1 - p$.

This method of making a decision about the quality of semi-conductors is a poor one as the chances of making mistakes are very large. Now check that you understand the reasoning by working out the probability of accepting a batch when $p = \frac{1}{3}$. Lectures on or reading about the binomial distribution of probabilities might throw further light on this problem.

How could the inspection system be improved? Many systems have been devised and documented but all rely on taking more than one item from the batch and testing it. Some systems call for taking a first sample of say 10 items, testing each one and making one of THREE decisions:

(1) Pass the batch if there are no defectives in it.
(2) Take a further sample of 10 if there is only one defective.
(3) Reject the batch if there is more than one defective.

This scheme is an example of a general principle of statistical practice. If a decision-maker is not satisfied that the data presented to him is SUFFICIENT for him to make a decision which is 'very probably' correct, he should request more data.

The Role of the Statistic in Uncertainty

The analysis of situations, where a sample containing more than one article is taken, obviously calls for a more sophisticated study. Problems such as independence, the choice of the statistic, and the sample size, which do not arise when a sample of size 1 is investigated, need to be recognised and analysed. The answer to the problem of sample size is complex and depends on the precision and accuracy required for a given problem, so specifying these is an important aspect of the problem and its analysis.

Let us take a particular example and investigate it in detail. Suppose an accountant wishes to find the proportion of a bank's customers who are overdrawn. He could get someone to go through every account, recording whether it was overdrawn or not, but that would take a long time and the problem of accuracy would arise.

He settles for a sample. How many accounts should he investigate? If too few his answer would be imprecise, if too many he would have wasted time. A sample of size 3 accounts would give him a possible answer of:

(1) 100 per cent overdrawn, if all 3 chosen accounts happened to be overdrawn.
(2) 67 per cent overdrawn, if 2 of the 3 chosen accounts happened to be overdrawn.
(3) 33 per cent overdrawn, if 1 was overdrawn.
(4) 0 per cent overdrawn, if none were overdrawn.

Whatever the TRUE proportion of overdrawn accounts was, a sample of size 3 could produce ANY of these four answers. Any of these four results could be obtained from a sample of size 3 whether the true proportion was as low as 1 per cent or as high as 99 per cent. Naturally, if the TRUE proportion was 99 per cent, the 100 per cent overdrawn answer would be more probable than the 0 per cent overdrawn answer, but the 0 per cent could conceivably occur. A sample of size 3 is bound to give an imprecise estimate of the true proportion.

Its accuracy is not much better. If the true proportion happened to be 80 per cent, the sample of size 3 would give one of four answers, but at best the closest answer would be 67 per cent, an underestimate by 13 per cent.

A sample of size 3, we have established, gives neither a very accurate nor a very precise estimate of the proportion we are trying to estimate.

However, the precision can be increased and measured by indulging in a fanciful piece of imagining.

Imagine, if you can, our accountant selecting 100 customer entries at random and recording the proportion in that sample of 100 that are overdrawn. This proportion is called the 'value of the statistic'. That first piece of imagining was easy; the more difficult piece is to imagine that he selects another 100 customer entries at random and records this proportion as well. Not content with a piece of apparently useless repetition he repeats the exercise of selecting 100 entries a very large number of times. Obviously his 'value of the statistic' will vary from sample to sample: it could, conceivably, range from a value of 0 to a value of 1 but it PROBABLY would not, the range of values being realistically limited to a much narrower region. Let us now assume that, unlike the struggling accountant, we know the TRUE proportion of overdrawn accounts to be $\frac{1}{3}$ or 0·33 or $33\frac{1}{3}$ per cent.

Plotting our statistic values as a histogram we obtain the PROBABILITY DISTRIBUTION OF THE STATISTIC. For 0·33 and a sample size 100 (the number of samples is irrelevant, fortunately!) this is given below.

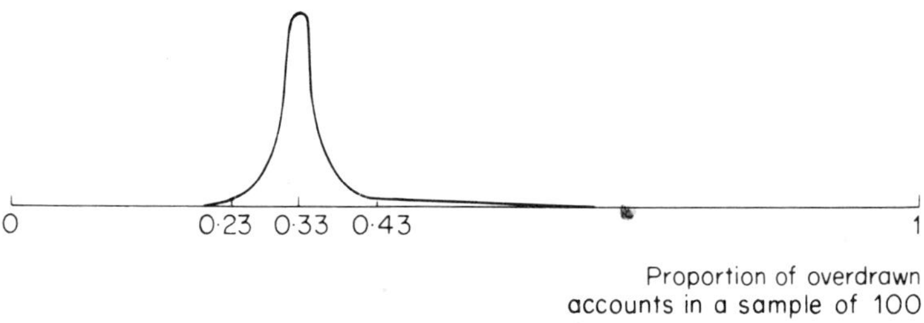

What it says is: If a sample of size 100 is taken from a set of accounts, $\frac{1}{3}$ of which are overdrawn, the proportion of overdrawn accounts in any ONE sample of 100 accounts will very likely be in the range 0·33 ± 0·10, OR nearly every sample will provide us with a proportion within 0·10 of the TRUE proportion. Take a sample of size 100 accounts and you can be 'pretty sure' that you will be within 10 per cent of the true proportion – providing that is about $\frac{1}{3}$. Interesting, you say, but where did the 0·10 come from and how sure is pretty sure?

We will discuss these points later, but meantime let us consider the wider implications. From any sample of measurements a number of statistics could be calculated. We have intuitively chosen the statistic, 'the proportion of overdrawn accounts in a sample of 100 accounts'. We could have chosen as our statistic, 'the number of overdrawn accounts in a sample of size 100' or 'the number of overdrawn accounts divided by the number in credit'.

The descriptions of these statistics is cumbersome and since the same one is used in many different situations they are given names. You can imagine the difficulties the early statisticians had in naming the first statistics by the names they gave them. They range from the mundane, $\bar{x}$ and s^2, through the esoteric χ^2 to the ironic *t(ea)*, given to an early statistic by a statistician who invented it while working for a famous brewery. The names of the statistics reflect the personality of their originators; if you don't like $\bar{x}$, for instance, you can privately think of it as Fred or Elsie.

We will now name our statistic. We name 'the proportion of overdrawn accounts in a sample size 100', *P*100. *P*100 is not a unique name for this statistic, other writers use P_{100} or just *P*. The name is immaterial, the definition is the important thing.

Now, the 'value of *P*100' would vary from sample to sample if many of size 100 were taken. Like all statistics, *P*100 will have certain properties: it will have a probability distribution which is most important, a mean and a standard deviation.

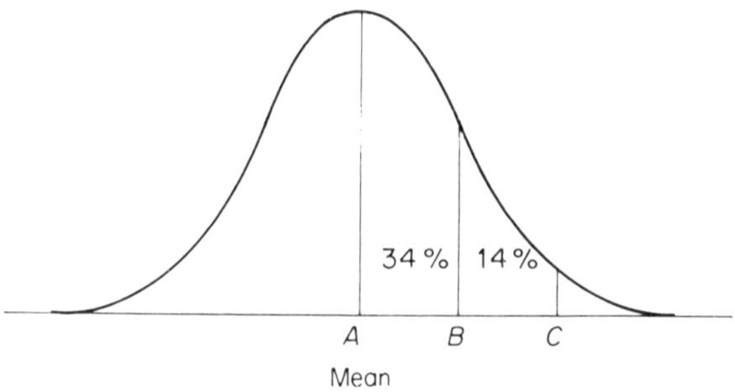

The probability distribution of the statistic P100

ALL statistics have probability distributions and names. Some are more popular than others because they have characteristics which enable general inferences to be made from the sample. In the case of *P*100 it has a mean *P*, the true proportion of overdrawn accounts, a standard deviation of $\sqrt{P(1-P)/100}$, and its probability distribution is normal.

Now we can use these characteristics to investigate the precision of *P*100.

The distance from *A* to *B* is one standard deviation $= \sqrt{\frac{1}{3} \cdot \frac{2}{3}/100} =$ about 105. The distance from *A* to *C* is two standard deviations: about 0·10. So 96 per cent of all the 'values of the statistic' will be in the range 0·33 ± 0·10. It will be very rare, probability about 0·04, to get a value outside the range $0{\cdot}33 \pm 0{\cdot}10$. $P100$ will have a precision sufficient to ensure that its 'statistic value' will be within 10 per cent of the true proportion.

Now we take the more realistic stance of being uncertain about the proportion of overdrawn accounts. There are at least two ways of approaching the problem and one is to use the data to modify your degrees of belief about the possible value of the proportion. Obviously, the argument runs, you would be inclined to favour a particular set of proportions rather than another set before you collected the data and then the data merely modifies your previous belief in the probable values of the proportion. Read Lindly, page 15, for further insight into this way of looking at things.

The other approach is to eschew subjective probabilities, pretend that they do not exist; the only hard fact is the data. Suppose that the statistic value, the proportion of overdrawn accounts calculated from the sample, is $0{\cdot}20$, and the sample size is still 100.

What do we know? First of all that the sampling distribution of $P100$ is normal. Its mean, which also happens to be the unknown proportion P, is what we wish to find.

Secondly, we know that the formula for the standard deviation of $P100$ is $\sqrt{P(1-P)/100}$. Unfortunately, we do not know P; that is what we want to find! How can we use what we want to find to find itself? That is just what we shall do.

How big is $P(1-P)/100$? P is between 0 and 1 and if P is large, $1-P$ must be small by comparison. If you plot $P(1-P)$ against P you will find that at $P=0$, $P(1-P)$ is 0 and it is 0 again when $P=1$. Between these two points $P(1-P)$ must reach a maximum.

Try your differentiation out on $P-P^2$ or plot the function $f(P) = P(1-P)$ for $P = 0$ to 1, and you will see that $P(1-P)$ never exceeds $\frac{1}{4}$ no matter what value the unknown P takes.

What have we now? The standard deviation of $P100$ is AT MOST $0{\cdot}5/\sqrt{100}$, about $0{\cdot}05$.

This means that our value of $P100$, $0{\cdot}20$, cannot be more than AT MOST $3 \times 0{\cdot}05$ from P and is unlikely to be further away from P by more than $0{\cdot}10$!

Answer: P is likely to be between $0{\cdot}10$ and $0{\cdot}30$.

How big a sample would the accountant have to take to be reasonably certain of being within $0{\cdot}01$ or 1 per cent of the true proportion? Work it out from the formula for the standard deviation of the statistic $P100$ with n replacing 100.

What we have discussed so far ranges over some of the statistical ideas which should prove useful to anyone whether in industry, research, education or the social services who has to manage resources effectively and make sensible, informed decisions on the basis of the information available at the time. But this is not the whole story. In the same way that probability is interpreted in different ways by the use people make of it, 'statistics' has many facets too.

Sherlock Holmes in the *Sign of Four* commented that when you have eliminated all the probable hypotheses by testing them against known facts, what you are left with, however improbable, must be the truth. This is the philosophy of the hypothesis tester, the sleuth, the psychologist, the deviser of new 'theories', but every statistics user must decide whether he NEEDS to test hypothesis. If you decide that hypothesis testing is your field then read statistics books full of these tests. Read a wide variety of books, just for their hypothesis-testing content, but be constantly on guard against the specialist approach to the subject by workers in a particular field, medicine or agriculture perhaps, who tend to view the subject in restricted terms. They know from experience that in their areas of interest certain ideas are unlikely to be useful so these are never discussed, though they might be very useful to you. The only solution is to read widely.

Some people hold that statistics of the hypothesis-testing kind is only useful during the birth and babyhood of subjects, when data is sketchy and theories are few on the ground; as the subject grows so statistics is used less and less. Ask yourself: 'Is the subject I am studying young and not yet sure of the direction it is taking, more in the question without answer stage, or is it well established with hundreds of laws which have been verified time and again?' The answer to your question may well decide the type of statistics you will need to study and the books you will skim through.

There is not ONE statistics, there are many. All talk of probabilities, statistics, probability distributions and you can learn about these techniques from most introductory texts. Choose one which appeals to you. There are upwards of a dozen to choose from.

The interpretation of the application of these techniques depends very much on the questions you ask and the prior information you have available. There is no simple way to learn interpretation. Follow one book or lecture course for a while and then have a change. Question, argue, discuss and, above all, read widely but selectively.

8
Be Sure of the Data

Questioning the Data

Decision-makers rely on data so they have to ensure that they use the best available. Other people are only too willing to supply information for the decision-maker to use. There is a lot of this information about but much of it is inaccurate and a considerable amount of the remainder is not relevant to the decision on hand.

Recently, for instance, there was an article in an educational research journal reporting on the effectiveness of tutorials. The report concluded that when lecturers did not attend a tutorial, students talked more.

If the educational decision-maker was to use this data in the planning of his future courses he would need to know a little more about it. He would begin by asking questions about the data.

HOW MANY? (this is usually a good opening question). How many tutorials were investigated? There were actually 13.

WAS THERE ENOUGH DATA TO JUSTIFY THE CONCLUSION? (if you don't take enough data the generality of the result must be in question). DO you think 13 was enough in this case?

IS THE DATA RELEVANT? (psychological research results based on experiments with rats, for instance, cannot be automatically applied to humans).

IS THE RESULT UNIVERSALLY APPLICABLE? Are the students in the study the same as the ones we meet? They were in fact male undergraduate New Guineans and Papuans. Are these typical of the students we know?

Simple questions like these will search out the shortcomings of the data on which statements are based and will either suggest modifications to the statement, taking account of the limitations of the data, or indicate whether a completely new set is required.

Read Darrel Huff's book, *How to lie with statistics*, which contains a wealth of good advice. Do not be put off by the humorous style; everything he discusses should be taken seriously.

Reflect on the following piece of data before you decide to buy your next packet of cigarettes: 'If you smoke over 15 cigarettes a day your chance of suffering from heart disease is doubled.'

Ask:

> WHO says so?
> WHAT are his qualifications?
> HOW BIG a sample was taken?
> WHAT brand of cigarettes?
> WHAT SORT of people were investigated?
> HOW LONG was the investigation?

The answers to these and other questions you might ask will affect the extent to which this data will modify your decision. What is left out of a statement is often as important as what is put in. It is most important not to read into a statement more than is actually there. For instance, if you put a quantitative interpretation on a statement intended to be read qualitatively, then you will be misled.

When assessing the usefulness of data presented to you, beware of qualitative assertions masquerading as quantitative statements. Words like always, every, empty, half-full, over-worked, thick, thin, wet, dry, are usually context dependent, they mean one thing to the person telling the tale and something quite different to the listener.

The word FULL is such a word. A trainee manager working at a warehouse was told that the warehouse could not take any more stock; it was FULL. To check this statement he went to the source of the data supporting the statement, and saw that the warehouse had space to spare – it was demonstrably not full. 'Ah,' said the warehouse supervisor, taking pity on the poor trainee who could not be expected to appreciate the intricacies of the warehouse system, 'the floor space is divided into sections according to the type of material we stock. (All stock came in standard boxes about 1 metre cube.) The floor space is now completely allocated, so the warehouse is FULL – we cannot accept any new types of material. We need a bigger warehouse!'

It became clear to the trainee that the statement he should have been presented with was: 'Under the PRESENT SYSTEM, the warehouse is full!' Change either the warehouse OR the system and the statement is no longer true.

To use quantitative rather than qualitative statements is a step in the right direction so far as accurate description is concerned, but there are also dangers in this approach. People quickly learn that calculating statistic values from data is child's play; tedious maybe, but not difficult. Some people, once they learn to calculate, never stop: they seem to find satisfaction in the routine manipulation of figures – irrespective of what they mean. Yet figures must have come from somewhere. They are disembodied symbols only to a very pure mathematician – never to a decision-maker.

Suppose, for instance, our dedicated calculator finds that the average weekly income in an industry such as engineering has risen by £5/week. This sounds a lot and our decision-maker could decide, on the basis of this data alone, that demands for a further pay award were unjustified. But a more mature manager would ask questions like:

What sort of average are we talking about: mean, median, mode, another type?
Was the £5/week due entirely to a pay rise?
Was the £5/week increase partly due to overtime working or a recognition of increased productivity?
Did everyone get a £5/week rise, or did a few men get a £1000/week rise?
What was the average wage relative to similar industries before the rise?
Was the period used in the average calculation, typical?

An average considered in isolation from the data is a poor basis for making decisions. Even students lectured to at college are warned to look at the standard deviation when they are presented with a mean.

Your standard deviation, unfortunately, is not a cast-iron defence against a misleading summary of data.

A worried ex-student telephoned his tutor.

'I have forgotten how to present a statistical report to higher management,' he confessed.

The tutor, knowing from personal experience that it is not easy to present technical information to a non-specialist, asked for the whole story, warts and all.

Apparently the student's department had acquired a computer terminal with access to a statistical package. The student had carefully recorded the times taken for the computer to respond to a typed statement on the terminal teletype. A thorough man, he had kept at his post for 3 days. Then he used the teletype to calculate the mean and standard deviation of the response time. The results of the analysis worried the student. The statistical package came up with the answers:

$$\text{mean response time} = 3{\cdot}00562 \text{ sec}$$
$$\text{standard deviation of response time} = 14{\cdot}5739 \text{ sec}$$

How, the student asked, was he to advise his supervisor on the presentation of this information to his superior. The short answer was: 'Don't present this information.' Mentally the tutor sketched out his impression of the two figures, without the superfluous decimals.

'How long do YOU usually have to wait for a response?' he asked.

'About a second.'

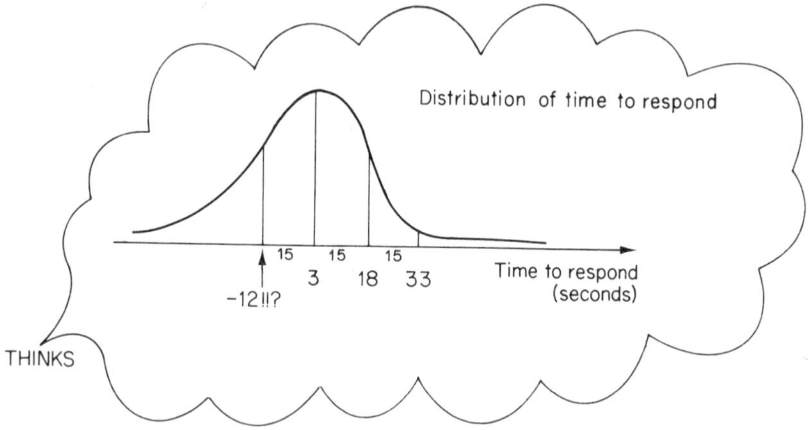

'And what is the longest you have waited?'

'About 4 seconds – unless the terminal breaks down!!!'

They then discussed the normal distribution, recalling that range was approximately 6 standard deviations, and decided that negative response times should not only be improbable but downright IMPOSSIBLE.

The student's report eventually recognised TWO types of operating conditions: normal and abnormal. Under normal conditions the response time was 'almost certain' to be within the range 0·5–4·5 sec with a mean response time of 2·5 sec. Occasionally, about twice a day on average, the terminal stopped behaving normally and was out of action for up to 60 minutes at a time.

This was the data which should never have been in the original calculation. It should not be ignored, the student's instinct was correct here, but it was not correct to include in the calculation data which turned out to be from a different source. Each problem should be carefully studied on its own merits to decide whether it is one problem or whether it is in fact an amalgamation of several quite separate problems. The appropriate data should then be collected for those parts and each problem should be solved individually and presented separately.

Collecting the Right Data

We now need to study the correct methods of collecting data to ensure that sensible statements can be made from it. Then we will be in a position to judge the validity of other people's data and the statements they subsequently make about it.

One of the mistakes people often make is to collect data for no specific purpose. This is a waste of time and money, but a lot of people still do it. The researcher who adds on to his questionnaire half a dozen extra questions 'because he thought the answers might be interesting' is wasting his and the interviewee's time. The stock controller who keeps inventory sheets up to 50 years old in the hope that 'they will come in useful – sometime' is wasting space and effort in keeping them filed. If data was collected only when there was a specific aim in view much more would be accomplished with much less effort.

BEFORE data is collected the use which will be made of it must be set down in writing. The decision or decisions which will be made on the basis of the data will affect the way in which it is collected.

At the simplest level, if we wished to decide whether middle-aged men living in semi-detached houses in the Midlands drank more beer than their opposite numbers in the south, from a general survey into the drinking habits of the population, we would have to ensure that there were enough men in each category to make the comparison valid.

People in samples (Midlands)

	Men			*Women*		
	Young	*Middle*	*Old*	*Young*	*Middle*	*Old*
Semi		○				
Detached						
Council						

Enough in this case might be at least 15, which could mean interviewing roughly 2 × 15 × number of categories in the table, which is 540 people. You could argue that there should be more or less, depending on the proportions of people in each category, but 540 gives a rough idea of the figure involved. It isn't 100 and it isn't 1000; it is somewhere between. And this calculation is usually sufficient to indicate whether it is economic to ask this sort of question in the survey.

A poorly thought out piece of data collection might well have started with the obscurely-worded objective: 'What are the beer-drinking habits of people in Britain?' This could not be the basis for a survey, it is too vague; too many decisions are left to the people collecting the data who will not be associated with its analysis.

It is no use asking detailed questions of the data AFTER it has been collected when the questions were not thought of BEFORE the survey was started.

Thought must also be given to the time scale of the enquiry, the cost, and the production of the data in a form suitable for the decision-maker to use. Many enquiries are started just to delay the taking of a decision and have little chance of success because no realistic appraisal is made of the resources necessary for an effective study and there is no intention of committing the necessary funds or manpower for the time the study needs.

Just what is involved in collecting data in a form suitable for the decision-maker?

(1) Decide on the time scale. Set a date for the final presentation of the information.
(2) Divide the time scale into sections:

  (*a*) Preparation, definition, preliminary investigation.
  (*b*) Design of the data-collection form in consultation with a statistician and, if the data is to be analysed on a computer, someone familiar with the form of input and output of the information.
  (*c*) Test the form on a few people – correct it.
  (*d*) Select the interviewees (people or objects).
  (*e*) Train investigators to collect the information.
  (*f*) Collect information.
  (*g*) Check that all the investigators did their jobs correctly.
  (*h*) Transfer the data into a form which can then be analysed, usually punched cards or magnetic tape.
  (*j*) Check that the transfer of data was carried out accurately.
  (*k*) Analyse the data.
  (*l*) Present the results.

By putting realistic times on each of these activities you will see that actually collecting the data represents a small proportion of the time needed to complete the investigation.

Sample Design in Business Research by W. E. Deming, published by Wiley, gives many good examples of the problems of collecting information, even though it is written from the statistician's point of view.

The Art of Asking Questions by S. L. Payne, published by Princeton University Press, is a difficult book to find except through a library, but it is well worth reading if only to remind you that many questions written on a questionnaire are ambiguous or misleading the first time they are written down.

We have mentioned several times in this book that data is usually incomplete and the information derived from it is only probably correct. The uncertainty associated with data cannot be eliminated completely but it can be reduced to an acceptable minimum level by collecting it correctly.

Only a correctly collected set of data will enable you to make a probability statement about a situation. For example, if an enquiry was conducted into the daily amount drawn on petty cash in a firm, a properly conducted sample should ensure that a statement can be made of the form: 'The average daily amount drawn on petty cash is £25 ± £5 and I am 95 per cent certain that this statement is true.'

It is only possible to express a degree of belief in a statement based on data when it has been collected in a probabilistic manner. In practice, this means that every daily amount drawn on petty cash must have a known probability of being included in the sample of data.

An elementary solution to this problem would be to write all the daily amounts drawn on similar slips of paper, drop them into a large container, shake them up and withdraw a given number blindfold. This method, assuming that the mixing was thorough, gives each slip an EQUAL chance of being selected for the sample. As such, this method is an example of a *probability sample:* a randomly selected sample with each member having a known probability of being selected. Equality of opportunity is NOT the criterion for selection. The only requirement is that the method of selection of units for the sample should ensure that each unit should have a *known*, but not necessarily *equal*, chance of inclusion in the sample. The apparently objective method of assigning can give demonstrably UNrepresentative samples. Let us see how this could arise. A teacher is asked to select four 'typical' engineering students from a class consisting of 8 women and 24 men. What should he do? Give each student an equal chance of being selected by writing their names on identical slips of paper and choosing 4 at random, blindfold? This method could produce a 'typical' sample consisting of 4 women, giving the impression that the typical engineer is female. Many samplers faced with this situation sense instinctively that this is wrong, discard this untypical sample and select another. But you cannot judge the validity of a sample by the result it produces. It is the method of selecting the sample which gives a sample its credentials. Taken to its logical conclusion, the sampler would continue sampling until he was satisfied that the sample he achieved was 'typical'. This is tantamount to selecting what he personally thinks is representative of the material he is sampling – it is not statistical sampling.

What has gone wrong with the reasoning? The method of allocating the same probability to each student by writing names on similar slips of paper was correct, so was the method of selection. How could we then get an 'obviously' incorrect answer? The reason is that all the information available was not used. We KNOW that there are TWO types of engineer – male and female. We should use this knowledge by dividing the class into two groups, male students and female students, and selecting some males and some females. Males and females are different in this problem, male engineers are likely to behave or react differently from female engineers. A 'typical' group of 4 engineers should include some of each type. The partitioning of a 'population' in this way is called 'stratification' and the selection of a sample involves a technique called stratified sampling.

The population in this situation is the class list from student number 1 to student number 32. This list is called the FRAME and consists of the numbers

1 2 3 4 5 6 7 8 9 10 11 . . . 29 30 31 32

Before attempting any sampling exercise the population must be listed and an unambiguous designation must be given to every member of it. This is not always easily achieved but it is an essential part of selecting the sample.

In this instance the frame is divided into two strata: one contains the 8 females and the other the 24 males. The strata are constructed on the basis of making members of the same strata as alike as possible:

The two strata

8 females	4 5 11 19 24 26 29 31	Stratum I
24 males	1 2 3 6 7 . . . 28 30 32	Stratum II

Logically we would select $\frac{1}{4}$ or $\frac{8}{32}$ of the sample to be female, the rest male. The single female would be selected at random from the list of numbers in Stratum I. We KNOW the probability of any female being selected from the sample and we KNOW the probability of any male being selected. They are equal in this case. They need not be but as long as they are KNOWN the sampling procedure is valid. For instance, if a survey was conducted to determine whether engineers preferred the classics to modern literature and it was known, from some other source, that male engineers all had the same unknown literary taste, this extra knowledge would be used in the sample. ONE male engineer would be selected for the sample since the information he gave would describe exactly the MALE attitude. This has carried stratification to its ultimate goal – to reduce the variation in a stratum to its lowest level, ZERO.

Put very simply, the more that is known about the population of values being sampled the smaller the sample size necessary to achieve the required precision.

Four was a very convenient number to choose for the sample size. How would we have allocated the sample between the two strata if the size had been 5 or 7? How do we choose the size of the sample anyway?

If the population can be divided into strata on the basis of information already in hand, then there are only two questions to ask in order to take an effective sample: how many are there in each stratum, and how variable are the values in each stratum?

Collecting Enough Data

If a firm were stock-taking, one way to proceed would be to divide the factory into low stock-value areas and high stock-value areas. Very roughly these two strata would correspond to the raw material stores and the despatch department. A histogram of the value of units within the factory might be as follows:

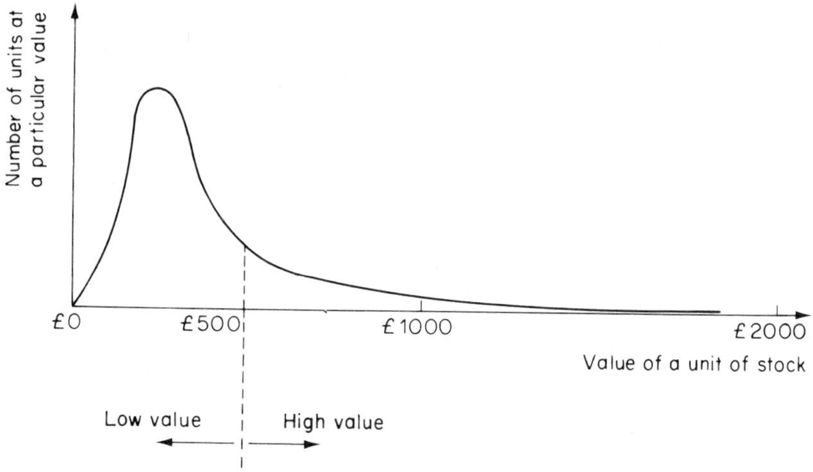

In the low-value stratum there are many units, with a mean value of about £250 and a standard deviation of about £500/6 (the range of a variable having the normal distribution of probabilities is about 6 standard deviations). But in the high-value stratum there are relatively few units, each having a high monetary value, and this high monetary value is itself very variable, from £500/unit to £2000/unit. Given this information we might well choose to determine the value of EVERY high-value unit, that is include the WHOLE of stratum II in our sample. Stratum I would be sampled sufficiently thoroughly to ensure that the average value per unit could be estimated to the required accuracy. Let us consider an actual case.

Suppose we wished to estimate the value of the stock correct to within $2\frac{1}{2}$ per cent of its total value. First of all we put an upper limit on the value of the stock. After discussion with the management, perusing last year's figures, and taking into account this year's trading, we might come up with the suggestion that the stock value will not exceed £3·6 million; $2\frac{1}{2}$ per cent of £3·6 million is

$$\frac{5 \times 3{,}600{,}000}{2 \times 100} = £90{,}000$$

Since there should be no error in evaluating the units of stock in stratum II, the only error involved in the sampling will occur in estimating the value of stock in stratum I. A unit in this example could be anything: a trolley full of radio chassis, a table covered in half-finished amplifiers, a box of resistors or a tray of nuts and bolts, as long as it was unambiguously designated as a unit and was given a number.

We have £90,000 worth of stock for our total error so if we estimate the average value of a unit of stock we must ensure that our estimate is within the range,

$$\text{TRUE average} \pm \frac{90{,}000}{N}$$

where N is the number of units of stock in stratum I. In this example we will suppose N to be 20,000 units so we must estimate the average value of a unit of stock to within ± £4·50. The estimate of the total value of the stock in stratum I will be 20,000 × average value of the units in the sample.

We are only going to take one sample, but prior to selecting that sample we must decide on the sample size. From previous chapters we know that taking different samples of the same size, say 1000 units, would provide different values for the statistic 'AVERAGE VALUE OF 1000 UNITS). This statistic must have a name, a mean, a standard deviation and a distribution. We have the name already and its mean is obviously the AVERAGE of the units in stratum I – it is the mean of ALL the means of samples of size 1000 and so it is the average value of all the units in stratum I. This average is what we are attempting to find to an accuracy of ±£4·5.

The choice of sample size will obviously affect the precision of the estimate so let us look at the two extreme possibilities. By taking repeated samples of size 1, the statistic values would be the actual values of the units in stratum I, so the distribution of the statistic values would be approximately normal with a mean of about £250 and a standard deviation of about £500/6 = £83.

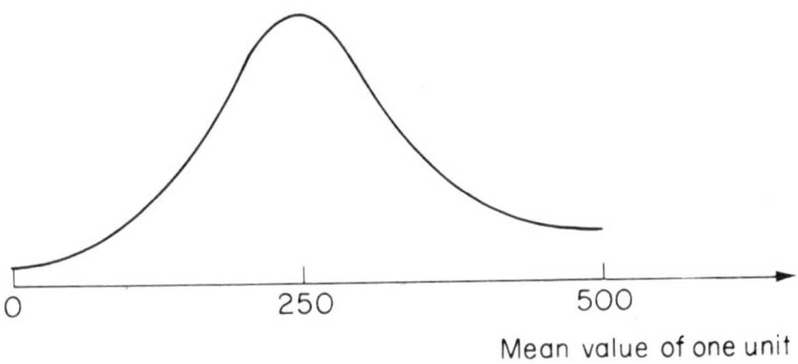

Distribution of samples of size 1

At the other extreme, by taking repeated samples of size 20,000 we would get the same mean for the statistic every time; there would be no variation.

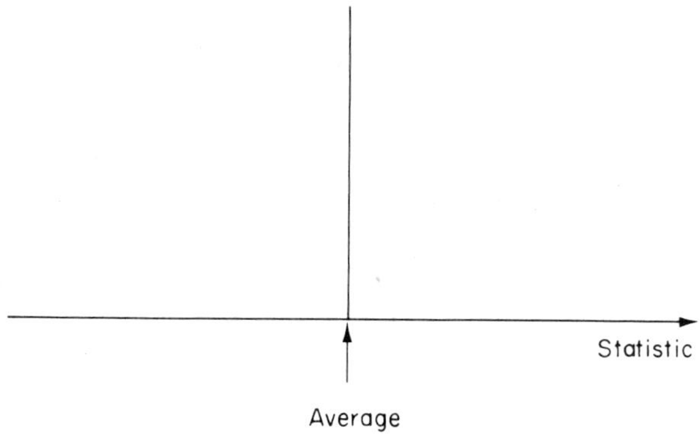

Distribution of means of sample size 20,000

The statistic with a sample size n, where n is between 1 and 20,000, will have a mean which is the AVERAGE we wish to find and a standard deviation somewhere between 0 and 83. By looking up the probability distribution, often called the sampling distribution of $\overline{X}$, the 'mean of a sample of size n' in an elementary statistics book, and applying this result to our example you will deduce that the standard deviation is $83/\sqrt{n}$. The distribution is normal.

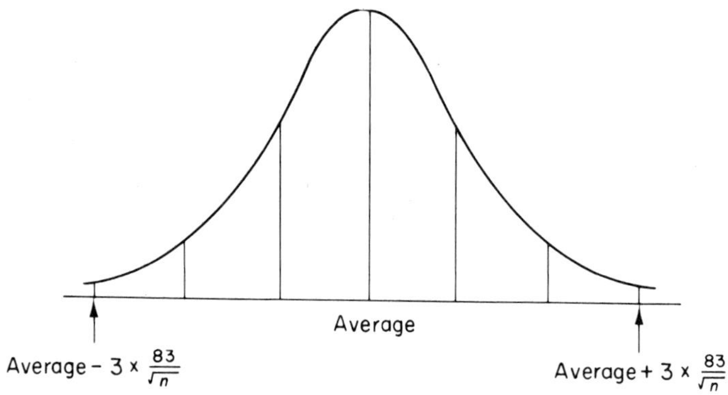

Distribution of the means of sample size n

Now we can determine n, the number of units, from stratum I needed to estimate the AVERAGE value of a unit. We need to ensure that the value taken by the statistic when the sample has been selected will be within ±£4·5 of the true AVERAGE. It is unlikely that the statistic value will be more than 3 of its standard deviations from the AVERAGE, so by adjusting the size of the sample n, we can make the standard deviation of the statistic, $83/\sqrt{n}$, as small as we like. We can make it so that the total sampling error we are allowing ourselves, £4·50, is equal to the TOTAL error the statistic can have, 3 standard deviations, so we put

$$3 \times \frac{£83}{\sqrt{n}} = £4{\cdot}5$$

and find that $\sqrt{n}$ is approximately 56 and n is $56^2 = 3136$.

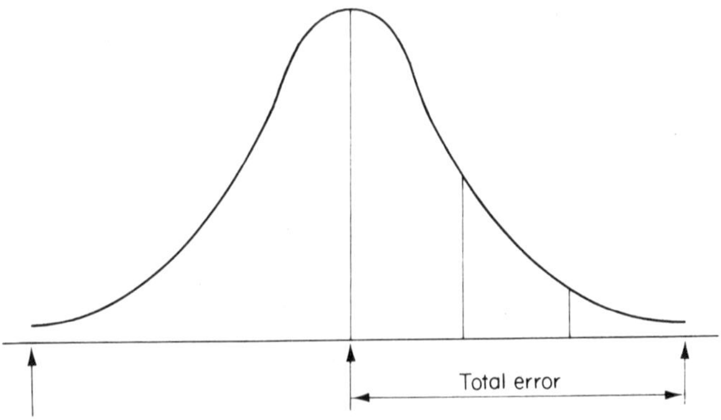

Total error due to sampling

By taking a sample of about 3200 units from stratum I and evaluating every unit of stock in stratum II we can be fairly sure that the stock value will be correct to within the $2\frac{1}{2}$ per cent we allowed ourselves.

The estimate of the total value of stock will now be

Total value = value of stock in stratum II + 20,000

× estimate of the mean value of a stock unit in stratum I

Now that we have worked through an example in some detail we can look back and recall important stages in the argument. The problem appeared to be to accurately determine the value of the stock in the factory, but what does accurately determine mean? Stocktaking results are used differently according to the decisions which will be based on them so we must ask: What is the reason for stocktaking and what decisions will be taken on the stocktaking result?

The reason might be to establish the proportion of the firm's capital which is tied up in stock, and the possible decisions might be to leave the stock levels as they are or to initiate a study into their reduction. For this purpose the firm wants to find out if the proportion is 10, 20, or 50 per cent correct to ±5 per cent so there would be little point in using a method to estimate the proportion which gave an answer correct to $\pm\frac{1}{2}$ per cent. To get an answer correct to ±5 per cent the firm does not have to value EVERY item of stock.

By considering these ideas we were led to redefine the problem for solution as 'estimate the stock value with reasonable precision'; consequently we decided that a properly selected sample would give us the answer. Any apparent increase we might have achieved by evaluating every unit of stock would be more than offset by the error incurred due to the employment of an army of unskilled stocktakers, and the uncertainty about the size of that error. By adopting a suitable sampling plan we would be able to allocate our resources to minimise the TOTAL error in the investigation and make an assessment of the probable size of that error.

Controlling the Collection

Taking samples is a skilled job. As a financial supervisor you would have to satisfy yourself that:

(1) The correct data was collected.
(2) Enough data was collected for meaningful decisions to be made.
(3) The sample was a probability sample.
(4) Any information about the population, available before the survey started, was used effectively, either to reduce the number of units sampled or to reduce the precision of the estimates.
(5) Enough financial resources are allocated to support the enquiry.

Now read the first 50 pages of A. Stuart's book, *Basic Ideas of Scientific Sampling*, Number 4 of Griffin's statistical monographs and courses, and a selection of the material contained in the first five chapters of W. E. Deming's excellent book, *Sample Design in Business Research.* Neither book requires much mathematics and each explains in straightforward terms the nature, power and techniques of sampling.

Using the Computer

Sooner or later in your career you will meet computers and their print-outs, so you must be prepared to deal with them effectively. First of all a computer calculation is only as good as the data which is fed into it; no amount of statistic-ing or computing can make bad data good. So, check the data that went into the computer, which usually means obtaining a print-out of the data which was processed. Failing this you should insist that the highest and lowest value of the data is printed out to check for mispunched or misread data.

Secondly, the program you are using should be checked out using simple data with which you are familiar. Putting in data containing deliberate errors will check how well the program deals with mis-punched data. Some programs do not recognise defective data and process it along with the correct data, with disastrous results. Thirdly, you should learn to tackle computer print-outs. Interpreting a print-out should be child's play – if you know what you are looking for; if you don't, you shouldn't be looking.

Print-outs usually contain a lot of information, some of which will be useful and some which can be safely ignored. The useful parts can be identified provided the analysis is understood in simple terms.

As a simple example let us consider an organisation which owns a chain of eating houses. When the organisation investigates possible sites for building a new eating house, it needs to know what the likely profits will be. A simple model of the situation might be summarised as

Profits = function of floor area and distance from the town centre

i.e. $P = f(x, y)$

$P = a + bx + cy$

where x is the variable: distance from the town centre in miles; and y is the variable: floor area in m^2.

If the values of a, b, and c, as yet unknown, could be determined by using data about profits, distance and area from houses already in the organisation then, providing x and y were known for a new site, the profits could be calculated by substitution into the formula

$$P = a + bx + cy$$

To be specific. If a, b and c had been found by a computer analysis using a multiple regression program, and their values were 100, −1000 and 0·3, the profits of a proposed site 0·33 miles from the town centre, with a floor area of 1200 m^2, would be found to be

$$P = 100 + (-1000) \times 0{\cdot}33 + 1200 \times 0{\cdot}3$$
$$= 130$$

Now we can ask some questions about the analysis:

(1) How precise was the 'a' value? 100 ± 0·01 or 100 ± 50?
(2) How precise was the estimate of P? 130 ± 20 or 130 ± 200?
(3) Is the equation, $P = a + bx + cy$, a good fit to the data? Would $P = a + bx$ or $P = a + cy$ be just as good?

(4) Would a curve, say $P = a + bx^2 + dy^2 + csy + ex + fy$, be better?
(5) Is there enough data to enable us to estimate a, b and c accurately enough for use in a prediction equation?
(6) How much of the variation in P is accounted for by $a + bx + cy$? 10, 30 or 90 per cent?
(7) How much variation should we reasonably expect to be explained by $a + bx + cy$? 90, 95 or 99 per cent?

If this information is not available on the print-out, ask for it to be included; if it cannot be included ask the management services department to produce it for you from the print-out. A bare print-out is rarely useful unless it supplies information that you need to use in making decisions, in a form that you can readily understand. If you need the data for specific decisions you will know what questions to ask; if you have no decisions in mind before the print-out is presented, you probably don't need to see the print-out at all.

The universal availability of computers has undoubtedly expanded the range of problems a firm might consider solving mathematically. Computer packages have brought many mathematical techniques such as simulation, multiple regression, linear programming, network analysis and stock control within the reach of managers who rarely used so much as a slide rule before.

All these techniques are very powerful, used properly, but the claims made for them have sometimes been out of all proportion to the use which could be made of them by the firm. When judging the usefulness of a technique, a manager should use the mathematical perspective we have tried to develop in this book. Just because facilities exist for carrying out complicated calculations on a computer, it does not follow that problems should be distorted to fit the computer package's requirements. As a matter of routine the manager should ask whether the technique is really appropriate to the problem, whether the data satisfies the assumptions the technique requires and whether the technique will provide answers to the problem with the required precision.

The mere fact that the answer is produced by a computer should not influence your judgement of the validity of the answer in any way. By doing a calculation on a computer you will not enhance either the data or the method and you should subject the answer produced by a computer to the same scrutiny you would use on an answer worked out by hand. A manager would not judge a written report by its weight and there is similarly no reason why he should be intimidated by large quantities of print-out. A print-out is, after all, only a numerical report, and the amount of paper it occupies is usually irrelevant so far as useful content is concerned.

Computer systems are often very successful at keeping up to date on stock levels. The stock level of every stock could be printed out weekly if required, together with details of stocks delivered in the last week and those on order. For a reasonably sized firm a print-out of the stock situation could be three or four inches thick and weigh several pounds; and no manager is going to successfully assimilate all the information contained in that print-out. He does not need to of course: his job is to check that the stock system is in control and to achieve this he can adopt one of several strategies.

If he is on good terms with the computer manager he can ask for an abstract of the movement of key stocks, confident that being aware of the condition of stocks he himself regards as important he can make decisions to query stock levels when the need arises. In another situation he might request just a list of the stock levels of items having high demand and those which haven't moved for a month or more. Usually only a small amount of stock-report print-outs is relevant to the decision-maker. He must know what to look for, record it if necessary on a graph to see its relation to previous weeks' values and be confident that in discarding the rest he is losing no important information.

Computers CAN amass large quantities of information but they should be encouraged to print out only that which is of direct relevance to the decision-maker. He needs easily assimilated data, which is seen in the context of similar data from previous weeks, displayed graphically if necessary, and data which indicates clearly irregularities such as extreme values.

The computer system as a management tool has no equal if the manager is in control of the system and not at the mercy of a computer manager who makes the decision by default about what should be printed out and what is worth calculating. A computer manager will provide a better service if you put him in the picture so that he can see that his output is used effectively. He would obviously prefer to produce two pages of special output for you, which he knows will be used, than a ton of standard output which is routinely consigned to the wastepaper bin. A manager should benefit from having a computer available but only if he knows what data he needs for his decisions and the form in which it should be presented.

9
Aftermaths

Be Mathematical

After the maths, now what? In this chapter we invite you to examine what you have achieved mathematically with the help of this book. We hope you have reached a higher level of numeracy and have gained some numerate skills. You may not be expert mathematicIANS but you can at least be mathematiCAL, in the same way that you can write without having to be a professional author.

You can be mathematical without having to worry about whether you can differentiate sine$^{-1}(x)$. Leave that to the mathematician; that's HIS problem. YOURS is to make sure that you know how the affairs under your control are progressing, to be prepared to take decisions about them and to know when and what decision to take. You should now be able to look at your problems mathematically and appreciate the advantages of this to your managerial role.

Be More Mathematical

The work presented in this text was selected specifically to help you develop mathematically. You should have found the topics themselves useful but now is the time to look at those not included here but which you might meet later.

The following list contains a few likely ones:

probability distribution theory
time series
index numbers
multiple regression
matrices
critical path analysis
linear programming
correlation
estimation
significance testing

The list is not exhaustive. It is not possible to make an exhaustive list of topics you may meet in the future; you cannot foresee the problem situations you may have to face, and you cannot anticipate the new mathematical ideas that will be developed in the next thirty or forty years. Your mathematical development does not end when your formal course of study ends: it is an ongoing process. Just as you may acquire a greater vocabulary the more books, newspapers and journals you read, so your mathematical appreciation and development will increase as you are exposed to more problems and to the techniques currently being discussed in the business world.

When you decide to learn something about a technique it is important that you adopt an effective strategy of study. You must first decide WHY you want to learn. Is it so that you will be able to use this technique yourself? If so, you must be prepared to spend considerable time with an advanced textbook, mastering all the intricacies. Or, do you want to know roughly what the technique is about, so that you can appreciate why it is being used and can communicate with the specialist who will be using it? This will require quite a different sort of reading: an outline of the method and a quick look at the sort of applications will probably be enough. Don't try to be an expert at everything.

Once you have established WHY you need to learn about something, you know HOW to do the reading. The next problem is to decide WHAT to read. It is most important to choose the book at the appropriate level for YOUR mathematical 'age'. For instance, if you know nothing about matrices, then your 'age' for this topic is ZERO. You may find that a book written for a 14-year-old, who also has a matrix age of zero, is very clear and conveys the elementary ideas you need more quickly than would an advanced text. As your matrix age increases you will find that you pose questions which the elementary book cannot answer; then is the time to consult a more advanced book. Reading an advanced book 'before your time' is unlikely to be helpful. Choose the book that is most likely to help you progress. It is impossible to make an exhaustive list of such books but in our experience these have proved very readable:

A. Battersby, *Mathematics in Management*
R. Langley, *Practical Statistics*
P. G. Moore, *Basic Operational Research*
W. W. Sawyer, *Vision in Elementary Mathematics*
M. J. Moroney, *Facts from Figures*
P. Rivett, *Principles of Model Building*
Reichmann, *Use and Abuse of Statistics*

Now try and find out what the topics in the earlier list are ABOUT.

When you see a book that might be useful, don't just look at the title; open it and look at a few pages to try to get an idea of the level and the method of presentation; look at the contents, introduction and index. It is worth spending some time searching for the right book because you will make faster progress when studying from it later on. If it isn't the right kind of book for YOU, put it down and find one that is. When you want to use it, don't try to read it like a novel and don't start at page 1: it is almost certain that you will have given up by page 3. Be clear about what you want to find out from the book, and what questions you require answering. Then select and read the relevant parts. It is often a good idea to read in addition the first and last chapters to put your reading into context (see, for example, M. J. Moroney's *Facts from Figures*). It is very unlikely that you will need to read the whole book; the chapters may be quite separate, particularly if it's a book on mathematics, and there may be little or no flow or development of ideas from beginning to end. Often a book is for reference, like a telephone directory, and you wouldn't read that!

Mathematical applications are not confined to books: we are surrounded by mathematical diagrams, analyses and statistical results in newspapers and magazines, on billboards and television. Try out your mathematical appreciation by criticising and analysing what you see and hear. Much will be misleading or incomprehensible, some blatantly incorrect, but even so you will pick up some useful ideas. In particular, study the newspapers for ways of presenting data on charts and diagrams.

You will also find mathematics and statistics used in other disciplines you may be studying, especially Economics and Accounting. Bring your mathematical judgement to bear on such areas as marginal analysis, breakdown analysis and stock control theory. If you are employed, find out the applications of mathematics in the firm. There will certainly be data collection and presentation, and statistical reports; you will find graphs of sales figures, tables of stock levels, an army of numbers that concern the day-to-day workings and long-term planning of the firm. The numerate manager must know how to react to the barrage of information confronting him.

In addition to this 'everyday' information, the manager must also know how to cope with 'once-off' reports and data connected with special studies. This may be more difficult for the manager to digest: the figures may be in a form which is unfamiliar to him and may be based on analyses he has never heard of. The extent of his understanding depends on the relationship he has built up with the person or department conducting the study through its many stages. Let us therefore consider the manager in his role of consulting a specialist.

The appropriate time to bring a specialist into an investigation is at the beginning. Consultations between him and the manager should take place at every stage: problem definition, data collection, analysis, solution and implementation.

The manager should make clear to the specialist the level of accuracy he requires and the cost, time and manpower constraints operative over the period of the investigation.

It is the job of the specialist to provide clear reports on the state of the project and to be prepared to elaborate on points that the manager wants clarifying. The manager is then in a position to fulfil his role of overall control.

Furthermore, the study itself will have increased his mathematical development, rather than leaving him bewildered, and he will approach new problems and their investigations with confidence rather than fear.

10
Examination of Examinations

Section 1

Examinations are usually set at the end of a course on mathematics. Sometimes they seem to have little relevance to the real world of business but the award of a useful qualification might depend on passing that examination. They should be treated like any other problem to be solved.

The problem must be recognised for what it is and DESCRIBED in detail. This description, which might be very personal and even emotional, must be ANALYSED and appropriate METHODS devised to solve it.

Section 2 of this chapter puts forward suggestions for your description, analysis and method of attack while Section 3 attempts to use actual examination questions which have been set in the past few years to illustrate examination technique.

You should work through the examples as you read them, plot the graphs, plot them again using different scales and make mistakes. Do the calculations using different codes. EXPERIMENT!

This book does not promise that mathematics can be made easy; mathematics is NOT easy. The only way to be successful in mathematics is to write it and talk about it.

Section 3, however, does provide the first few steps in the learning process. You are invited to tread them until you have enough mathematical confidence to tackle the new problems of YOUR examination.

Section 2

Examinations must be approached in a positive frame of mind. Remember that the examination is as much a test of the examiner as the examined. In fact, the examiner is hedged around with restrictions which leave him very little room to manoeuvre so that the chance of a unique type of question being set is very small. Just consider for a moment the examiner's constraints:

(*a*) Quantitative methods is often a subsidiary subject and a large failure rate in a subsidiary subject is unacceptable to an examination committee.
(*b*) The standard routine calculations must be included to some extent, and these are very easy to answer.
(*c*) A flavouring of descriptive sections is usually included to balance the pure calculation side. These sections are easily detached from past papers and if necessary learned parrot fashion.
(*d*) The syllabus is very limited and limits the choice of topic.
(*e*) It is not possible to devise a standard paper to prevent students passing by question spotting.
(*f*) He is only human.

Examinations are a problem like any other, and problems succumb to the correct strategy for tackling them.

Definition

Ask yourself

What is an examination?
What is it for?
What do I fear about it?
Who sets it?
Why do they set it?
How do they set it?
How would I set it?

Examinations mean different things to different people. Examiners would like to think that students approach examinations with the attitude: 'Now I have the opportunity to demonstrate my vast knowledge of the subject matter imported to me so brilliantly by my mentors.' Some students approach examinations thinking: 'This is just another way *they* have devised to demonstrate my inferiority.' Writing down your own attitude to examinations is a healthy exercise which is made even more valuable if it can be modified, by answering the questions above, into a more positive and determined approach.

Some of you reading this will wonder what all the fuss is about because you have always found examinations 'easy' and propose to think no more about strategies. It could be a mistaken view. Why have you found examinations easy in the past? Because you understood what they were all about and adopted a strategy appropriate to each one, or because you were lucky in spotting a few topics, or was it because you were too intelligent for the level of the examination. If the latter, you might recall that intelligence has a nasty habit of 'peaking out' during a high-level course, and survival will then depend on cunning, and application to the problems on hand.

Analysis

Analysis will depend on the answers you find to the questions proposed in the definition section.

Method

(1) Look at old examination questions.
(2) Ask your lecturer for definitions and written explanations of areas which frequently occur on past papers.
(3) Become an expert in two or three topics by selectively reading textbooks not recommended for the course, and pestering the lecturer.
(4) Get a syllabus and tie it in with the set book so that you can pick your topics before the lecturer reaches them in the course. (Do this today!)
(5) Be aggressive when you see a poorly set question. Redefine it in your own terms and offer a solution pointing out that the question as set is open to different interpretations.
(6) Be confident. This is usually a product of being well prepared by carrying through a thought-out strategy, but it might mean forcing yourself to lead seminar discussions and to admit to your lecturer that you do not understand some topic at some time during the course.
(7) Look through the examination questions and commentaries following.

Section 3

Correlation

Correlation is a favourite examination topic because it can be tackled in so many different ways. Some examiners just require the formula to be memorised and the calculation carried out, while others ask for detailed assumptions and interpretations. No matter what the requirements of the question on correlation are, the answer can be given by keeping the following steps in mind:

PLOT the values. This is always worth doing, because if the points should happen to fall on a definite curve the correlation coefficient is NOT an appropriate measure, and having said so there is no point in carrying out a calculation. (For examination purposes, of course, it is advisable to go through the motions of a model solution!)

CALCULATE. This can be made less painful if a suitable CODE is used and you can RECALL the correct formula.

INTERPRET. Some examiners ask if the coefficient is SIGNIFICANTly different from 0 while others insist on a CONFIDENCE INTERVAL, but whether they do or not you would be well advised to SQUARE your coefficient to bring it down to size and to give you a measure of its *usefulness*. The square of ρ is the percentage of variation in one variable explained by the other.

We shall consider these in more detail using actual questions.

PLOT

In the following question a curvilinear relationship would be expected since population growth is unlikely to be linear.

Calculate the correlation coefficient for the following data

Year	1961	1962	1963	1964	1965	1966	1967	1968	1969	1970
Pop^n in millions	52·8	53·1	53·6	54·0	54·4	54·6	55·0	55·3	55·5	55·7

The PLOT shows a definite if slight curve and there is no doubt that there is a relationship but the correlation coefficient is *not* the appropriate measure of that relationship, because the relationship is not linear.

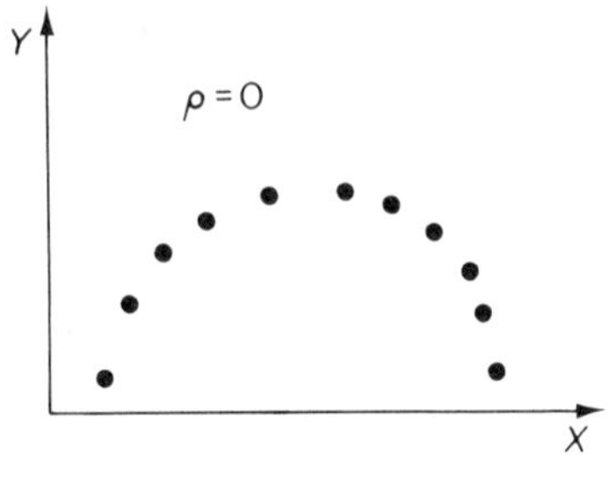

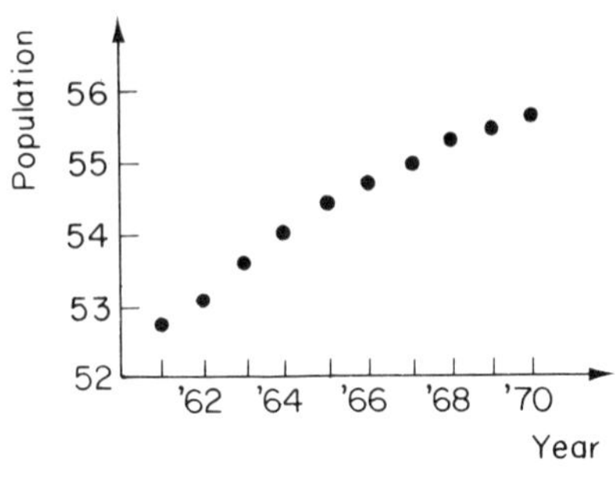

The correlation coefficient is only useful for measuring the degree of association when that association is LINEAR. If a correlation coefficient was the only criterion for measuring the degree of association between variables, then X and Y would be judged to be unrelated, because the correlation between the variables is ZERO. It is obvious from the diagram, however, that there is an almost perfect QUADRATIC association.

CALCULATE

The simple calculation problem is straightforward if unimaginative, but it is still worthwhile to plot the values, in the hope of injecting some controversy into the solution by suggesting non-linearity.

A plot will at least check that your calculated value of 0·99 for the correlation coefficient did not arise from data which looks like that in the figure.

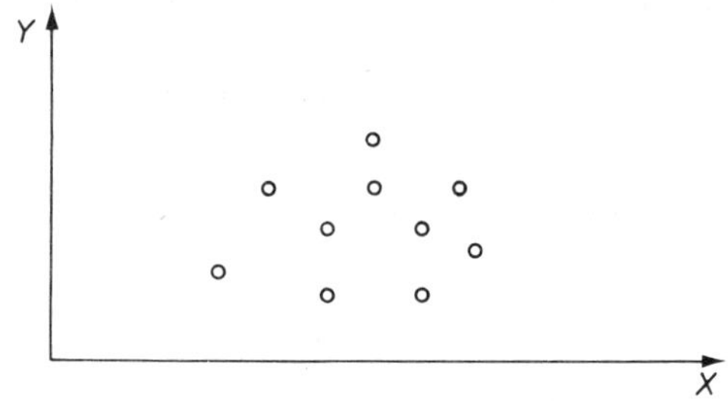

Coding will ease the burden of calculation and does not affect the value of the correlation coefficient. Work through the calculation below:

x	y	$X = x - 100$	$Y = y - 1300$	XY	X^2	Y^2
100	1298	0	−2	0	0	4
101	1297	1	−3	−3	1	9
102	1299	2	−1	−2	4	1
		3	−6	−5	5	14

$$r = \frac{3(-5) - 3(-6)}{\sqrt{\{3.5 - (3)^2\}\{3.14 - (-6)^2\}}}$$

$$= \frac{3}{\sqrt{\{6\}\{6\}}} = \tfrac{1}{2}$$

Substitute the following values calculated from the original data into the formula:

$\Sigma x = 303$
$\Sigma y = 3894$
$\Sigma xy = 393{,}295$
$\Sigma x^2 = 30{,}605$
$\Sigma y^2 = 5{,}051{,}816$
$(\Sigma x)^2 = 91{,}809$
$(\Sigma y)^2 = 15{,}163{,}236$

INTERPRETATION

Some examiners ask only whether a calculated value of the correlation coefficient is significantly different from 0. The question asked is just, 'is the calculated value of r significant at the 5 per cent level?'.

Essentially what they are only asking is whether there is the slightest chance that the variables are linearly related. The following question is a good example:

A works manager's weekly records of minor working faults show the following percentage of 'machine hours' lost, by the age of the machinery:

Age of machine (years)	1	3	4	6	8	9	11	14
Machine hours lost (%)	1·4	2·1	5·8	4·7	4·1	5·2	9·6	8·7

Calculate the product moment correlation coefficient for this data. Is it significant at the 5 per cent level?

You should take the opportunity whenever possible to augment your answer to such a question. Demonstrate your contempt for the inadequacy of the mere assertion,

'the correlation coefficient is significant'

by branching out and airing your specialised knowledge of the type of data presented for analysis.

In answering this question, for instance, ask yourself 'why was this analysis carried out?'

Was it designed to present a case for scrapping obsolete machinery or to show that older machines should be serviced more regularly?

Significance will have different interpretations depending on the reason for and the circumstances surrounding the collection of data. There could be, for instance, a 'significant' association between two variables as calculated by the correlation coefficient, but the information would be useless if the association was curvilinear!

QUESTION

The index of average earnings of all employees in Great Britain is shown below against the sale of National Savings Certificates for the same dates. Calculate the product-moment coefficient of correlation.

Month	*Oct*	*Nov*	*Dec*	*Jan*	*Feb*	*Mar*	*Apr*	*May*
Index	104	104	102	103	104	102	106	106
Sales (£ × 10^6)	30	23	16	28	24	19	47	40

SOLUTION

The plot indicates that the relationship could be linear. There is no evidence that it is curvilinear so a correlation coefficient can reasonably be calculated.

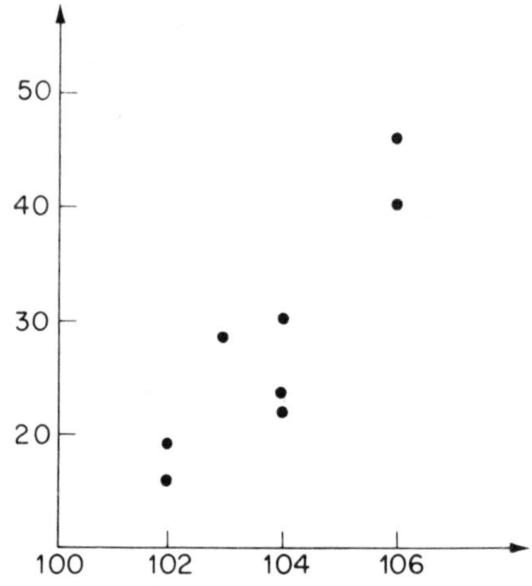

x	y	$x-100$ X	$y-30$ Y	XY	X^2	Y^2
104	30	4	0	0	16	0
104	23	4	−7	−28	16	49
102	16	2	−14	−28	4	196
103	28	3	−2	−6	9	4
104	24	4	−6	−24	10	36
102	19	2	−11	−22	4	121
106	47	6	17	102	36	289
106	40	6	10	60	36	100
		31	27 − 40	162 − 108		

$$r = \frac{8(\Sigma XY) - (\Sigma X)(\Sigma Y)}{\sqrt{\{[8(\Sigma X^2) - (\Sigma X)(\Sigma X)][8(\Sigma Y^2) - (\Sigma Y)(\Sigma Y)]\}}} = 0{\cdot}91$$

A very reasonable value for the data plotted on the graph.

A more sophisticated question type asks for more information.

QUESTION

What do you understand by correlation?

A works manager's weekly records of minor working faults show the following percentage of 'machine-hours' lost by the age of the machinery:

Age of machines (years)	1	3	4	6	8	9	11	14
Machine hours lost (%)	1·4	2·1	5·8	4·7	4·1	5·2	9·6	8·7

Calculate the product-moment correlation coefficient for this data. Is it significant at the 5 per cent level?

SOLUTION

First the plot.

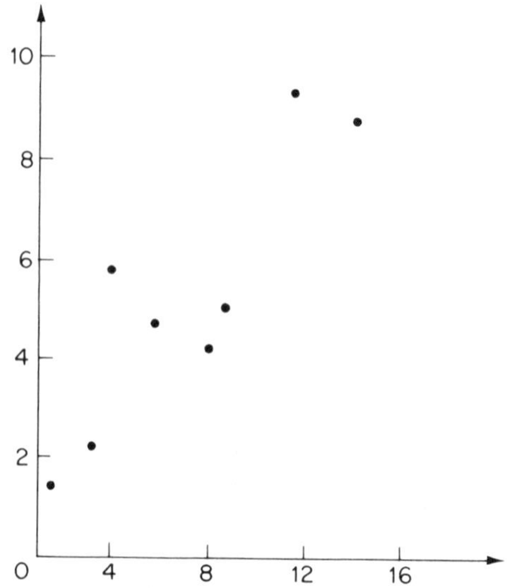

There could well be a linear association on the evidence of the plot.

The correlation coefficient is about 0·8 and there is very little doubt that it is 'significantly' different from 0.

Actual calculation gives $r = 0{\cdot}8598$.

The critical value for $r5\% = 0{\cdot}72$.

Comment. If there really was no linear association between the variables, samples of size 8 would give values of r less than 0·72, 95 per cent of the time.

A sample of size 8 from a population with $\rho = 0$ could *produce* a value of 0·8598 but it is very unlikely (less than 0·05), so it is more likely to have come from a population with a ρ value greater than 0.

The graphical approach is obviously appreciated by the examiner who set the following question.

QUESTION

(*a*) *For the four scatter diagrams given below discuss the value of the product-moment coefficient of correlation:*

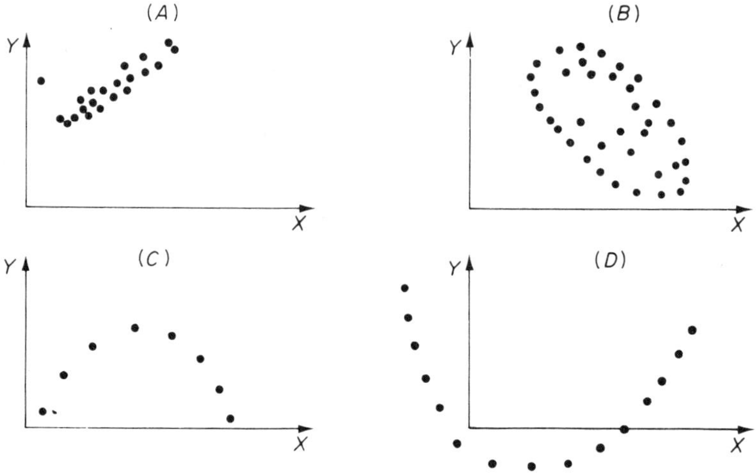

(*b*) *'The entrance test results in Mathematics and the final Mathematics examination results for 500 students gave a coefficient of correlation of −0·3.'* *Comment briefly.*

SOLUTION

(*a*) What is wanted here can be summarised as below:

(*A*) A strong positive correlation is indicated of about 0·9. *X* could be used to predict value of *Y* with a considerable amount of confidence.

(*B*) There is a weaker negative correlation of about −0·6. Predicted values would not be so reliable.

(*C*) The association is not linear so a correlation coefficient is not an appropriate measure.

(*D*) Same as (*C*).

(*b*) The 95 per cent confidence limits are approx. −0·37 to −0·22. There is a definite but very weak negative relationship between the variables. We could go further and square −0·3 to give us the measure of how useful the entrance test results would be for predicting final examination results:

$$(-0{\cdot}3)^2 = 0{\cdot}09$$

So 9 per cent of the variation in the final examination results could possibly be explained by the entrance examination results. This could reflect on the suitability of the entrance exam or the brilliance of the teaching!

This idea of the 'amount of variation explained' is asked for explicitly in the following question:

QUESTION

(*a*) *Calculate the product-movement coefficient of correlation for the following data on tensile strength and hardness of a sample of aluminium bars.*

Tensile strength (*000 lbs/sq. inch*)	*Hardness* (*rock wellt*)
2·9	53
3·5	69
3·8	83
3·0	55
3·5	76
3·1	64
3·6	70
3·1	53
3·2	80
3·3	67

(*b*) *What percentage of the total variation is left unexplained by the above relationship?*

SOLUTION

First the plot:

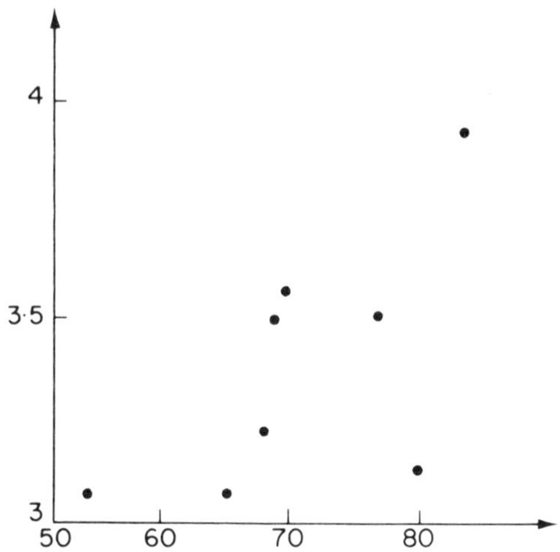

The calculation is made easier by coding (tensile strength −3) and (hardness −70). $r = 0{\cdot}787$.

95 per cent confidence limits for ρ are approx. 0·33–0·93.

Therefore the percentage of variation left unexplained is between 90 and 15 per cent.

In the next question interpretation is all.

QUESTION

(i) *Criticise the statements:*

(*a*) *A correlation coefficient of 0 indicates that there is no association between the variables under consideration.*

(*b*) *A correlation coefficient of over 0·8 almost certainly indicates that there is a causal relationship between the two variables.*

(ii) *There is a negative value for the correlation coefficient between open park space and the number of accidents to children caused by motor vehicles.*

What conclusions do you draw?

SOLUTION

(i) (*a*) We have seen from an earlier question that a correlation coefficient of 0 can arise from the situation in the sketch.

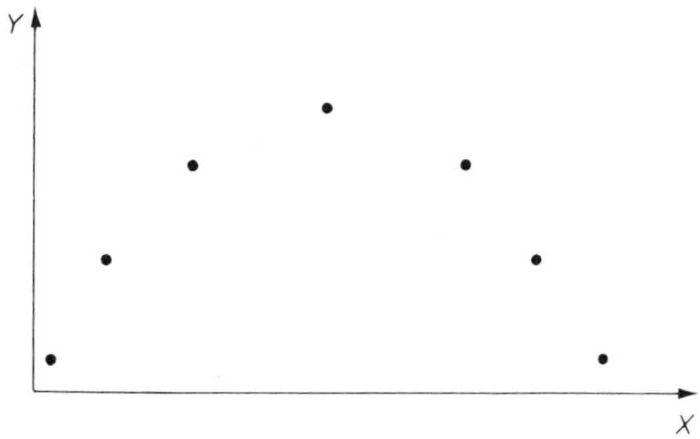

Here there is an association, but it is NOT LINEAR.

The missing word in the question's statement was LINEAR, i.e. ... no LINEAR association ...

The counter example in the sketch completes the criticism.

(*b*) The value 0·8 needs to be taken in context. Its interpretation depends on the number of values observed in the first place and obviously no amount of calculation could possibly establish whether the association indicated was causal.

(ii) There is a tendency for areas with large open park space to have fewer accidents to children.

We cannot conclude that creating open park space will cut down accidents to children, in the same way that we could not conclude that keeping children away from traffic would create open park spaces! Read *Facts From Figures* by Moroney (Pelican) for a full discussion.

So far the correlation coefficient has been used by the examiners to establish whether the association between two variables is linear. Once the linear nature of the relationship has been established it is natural to ask for the relationship to be stated.

If the relationship is linear then the two variables will be connected by the equation

$$y = a + bx$$

The problem is to find a pair of values, a and b, defining a line which in some way is a 'good fit' to the data.

Once a relationship is defined in algebraic terms it can be interpreted to help in the formulation of policies.

QUESTION

A firm wished to obtain the demand function for its product; the price and sales of this product for the past 9 years are given below:

	Price (£)	*Sales* (000)
1960	79	53
1961	74	62
1962	70	65
1963	79	54
1964	83	50
1965	86	48
1966	86	47
1967	89	46
1968	92	43

(*a*) *Calculate the demand function assuming it is linear.*
(*b*) *Calculate the price elasticity of demand at a price of £84.*

SOLUTION

The first problem is to determine which is the independent variable. Presumably in this case the sales depend on price. Fix a price and the sales will vary about an average sales figure for that price. It would hardly be possible to fix the sales and observe the price changes!

The plot, of course, is essential for two reasons:

(*a*) to see that the invitation to calculate a straight line is reasonable;
(*b*) to check that the calculated line does fit the data. Arithmetic errors are not unknown!

Coded values make the calculation easier:

X $x-80$	Y $y-50$	XY	X^2
−1	3	−3	1
−6	12	−72	36
−10	15	−150	100
−1	6	−6	1
3	0	0	9
6	−2	−12	36
6	−3	−18	36
9	−4	−36	81
12	−7	−84	144
36 − 18	36 − 16	−381	444

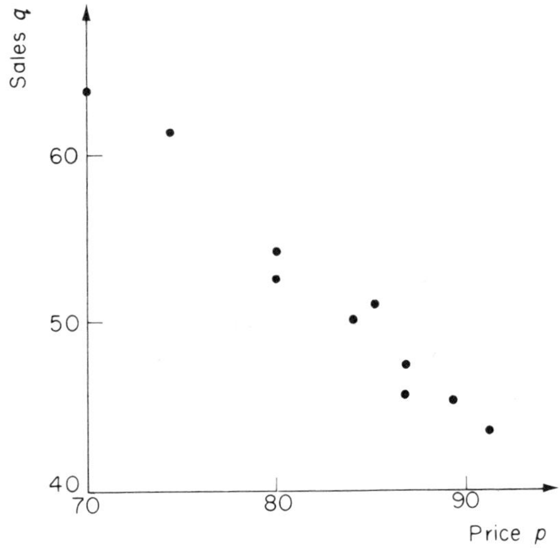

The slope is given by:

$$\frac{9 \cdot (-381) - 12 \cdot 20}{9 \cdot 444 - 324} = -1{\cdot}017$$

The slope is not affected by the subtraction of constants 80 and 50 from the original date.

The demand function is:

$$\text{Sales} = -1{\cdot}017 \text{ price} + 135{\cdot}402$$

The price elasticity at a price of £84 is η where

$$\eta = -\frac{p\,dq}{q\,dp}$$

$p = 84$

$q = 134{\cdot}402 - (1{\cdot}017)\,84 = 50$

$\dfrac{dq}{dp} = -1{\cdot}017$

$\eta = 1{\cdot}7$

The next step is a big one.

Having fitted a straight line to a set of values, some questions invite you to assume that the relationship will continue to be linear in the future. It is as well to pause and consider that the data for analysis is only part of a much larger picture, and that picture could take many different forms.

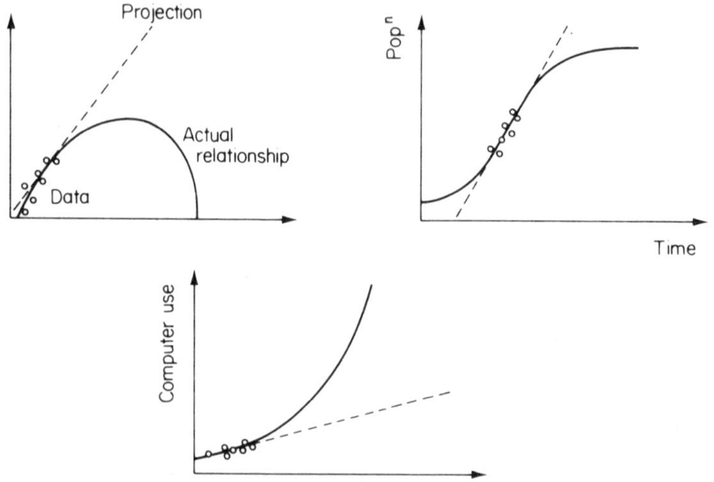

The decision to extrapolate cannot be taken only on the evidence of the values: the context of the data must also be considered. If necessary, several models should be fitted to the data and the answer given as a range of possible alternatives.

Consider the next question.

QUESTION

Fit a line of least squares to the following data and estimate the population of the U.K. in 1980.

Year	*Population (millions)*
1961	52·8
1962	53·1
1963	53·6
1964	54·0
1965	54·4
1966	54·6
1967	55·0
1968	55·3
1969	55·5
1970	55·7

SOLUTION

Plot the values with different scales for the population! The best straight line through the values can be more easily calculated if the data is coded.

x	y	$X=x-1965$	$Y=y-52$	XY	X^2
1961	52·8	−4	0·8	−3·2	16
1962	53·1	−3	1·1	−3·3	9
1963	53·6	−2	1·6	−3·2	4
1964	54·0	−1	2·0	−2·0	1
1965	54·4	0	2·4	0	0
1966	54·6	1	2·6	2·6	1
1967	55·0	2	3·0	6·0	4
1968	55·3	3	3·3	9·9	9
1969	55·5	4	3·5	14·0	16
1970	55·7	5	3·7	18·5	25
		5	24·0	51·0 − 11·7 = 39·3	85

$$b = \frac{10(39\cdot3) - 5 \cdot 24}{10 \cdot 85 - 25} = 0\cdot33$$

$$a = \bar{Y} - b\bar{X} = 2\cdot4 - 0\cdot165 = 2\cdot235$$

1980 will correspond to an X value of 15. Substituting $X = 15$ in $Y = 2\cdot235 + 0\cdot33X$ gives 7·28. This, added to 52, produces the 1980 estimate as 59·28 million people.

Descriptive Statistics

Raw data, that is values which are recorded as they are collected, is especially useful if someone else has summarised it into a form which is not suitable for your purposes. It can also give an indication of the accuracy of the investigator who collected the data. A set of data which gives values recorded to the nearest ½ mm would not be adequately summarised if the means were to be quoted to 0·000001 mm. Sometimes the only check on summarised values is to go back to the original data.

Raw data is, however, bulky and conceals, rather than points to, the essential features of the variable under study, so the data must be summarised in such a way as to describe the population of values from which the data was selected. Essentially, we need to know the SHAPE of the probability distribution, whether it is symmetrical or skew, where it PEAKS and how SPREAD out the values are. Knowledge of these essential features enables us to compare the population under study with other populations.

Often, fairly rough estimates of these three descriptive measures will enable valid comparisons to be made. However, it is not difficult in some courses to come to believe that the calculation of a standard deviation (preferably from raw data which has to be grouped, coded, and then corrected for grouping) is the initiation rite which will lead in some mysterious way to an automatic understanding of the art of data analysis. The effort required for this performance is so great in many cases that it can seem that the effort itself (whether accurately directed or not) should be enough to guarantee 100 per cent on any examination question. Unfortunately, effort alone does not guarantee success.

Examination questions asking for straightforward calculations of means and standard deviations are easy if somewhat tedious to tackle, and carry with them a time penalty, particularly if a poor coding scheme is adopted. If there is a choice on the paper, a question requiring thought can usually be despatched more quickly. If you do have time to tackle such a question at least check that your calculation value for the standard deviation is between $\frac{1}{3}$ and $\frac{1}{5}$ of the range and that the standard deviation is quoted together with its UNITS. Data about the resistance of components should not be summarised by quoting the standard deviation as 1·3056. It is quite a useless figure for our electrically-minded friend who wants to know whether it is ohms or megohms and does not want to have to go back to the original data to resolve the problem of units. It rather defeats the purpose of the exercise!

A typical run-of-the-mill question follows:

QUESTION

The following table gives the observed life in hours of 60 components of a particular type:

17	29	10	81	58	64	22	51	66	36	62	39
22	58	58	60	41	50	71	25	68	43	55	64
59	53	47	3	43	77	40	48	77	49	51	83
33	36	64	73	91	38	37	54	82	58	60	72
60	47	53	54	59	60	62	45	70	14	57	74

Tabulate the above data as a grouped frequency distribution, construct a histogram to display the distribution. Calculate the mean and standard deviation of the life of this type of component.

SOLUTION

The mean is about 50 hours.

The range is from 3 to 91, so an approximate standard deviation would be $90/5 = 18$ hours.

Boundaries	*Tallies*	*X*	*Frequency*	$x=(X-5\cdot5)/10$	*Fx*	*Fx*^{2}
0·5–10·5	||	5·5	2	0	0	0
10·5–20·5	||	15·5	2	1	2	2
20·5–30·5	||||	25·5	4	2	8	16
30·5–40·5	卌 ||	35·5	7	3	21	36
40·5–50·5	卌 卌	45·5	10	4	40	160
50·5–60·5	卌 卌 卌 ||	55·5	17	5	85	425
60·5–70·5	卌 |||	65·5	9	6	54	324
70·5–80·5	卌 |	75·5	5	7	35	245
80·5–90·5	|||	85·5	3	8	24	192
90·5–100·5	|	95·5	1	9	9	81
			60		278	1481

$$\text{mean of } x = \frac{278}{60} = 4\cdot63$$

$$\text{mean of } X = 51\cdot83 \text{ hours}$$

$$S.D.\ of\ X = 10\sqrt{\frac{1481}{60} - \left(\frac{278}{60}\right)^2}$$

$$= 10\sqrt{(24\cdot7 - 21\cdot4)}$$

$$= 10\sqrt{3\cdot3} = 18\cdot2 \text{ hours}$$

Note: $(5 \times 18\cdot2) = 91$

Note that the boundaries have been recorded to one decimal place more than the data; this ensures that each data value falls inside one and only one interval and there is no problem about a value falling ON a boundary. The boundary width is a convenient 10 hours; widths of 3, 7 or 2·65 are confusing. Choice of the lowest value 0·5 and 100·5 are not completely arbitrary but, since little guidance is given in the question, values which make calculation easy are selected. In a realistic problem you would have to use your own judgement about setting an upper limit; in this question the lower limit is about 0 hours so a limit of 0·5 hours will be a realistic approximation.

When the data is organised into equal intervals the coding is easy. The midpoint of the first interval is subtracted from each midpoint in turn and the result divided by the class interval. The resulting values always start at 0 and increase by 1 each time as the calculations proceed.

$$\text{Coded value} = \frac{\text{original value} - \text{lowest midpoint}}{\text{interval size}}$$

The mean of the original values is calculated from the coded mean as follows:

$$\text{Coded mean} = \frac{\text{original mean} - 5\cdot5}{10}$$

$$\text{Mean} = 46\cdot3 + 5\cdot5 = 51\cdot8 \text{ hours}$$

Why doesn't 5·5 figure in the calculation of the standard deviation?

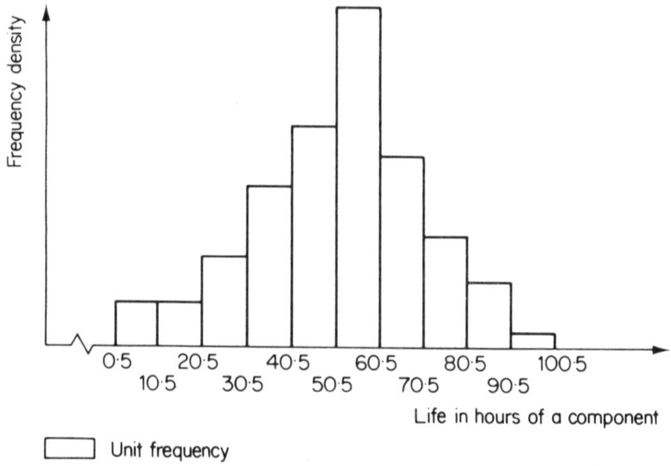

Histogram of the life in hours of a certain component
Source: Examination question

The histogram must stand on its own with no implied reference to any other source so it must have a title, clearly labelled axes, and a source for reference.

This next question allows for real analytical flair.

QUESTION

Samples of 50 of two brands of tyres A and B are taken and their lifetimes in '000's miles are as follows:

A	
'000's m	*f*
12 < 13	5
13 < 14	11
14 < 15	19
15 < 16	9
16 < 17	6

B	
'000's m	*f*
10 < 12	7
12 < 14	7
14 < 16	12
16 < 18	18
18 < 20	6

Compare these two distributions and decide which is the 'better' type. Choose your own criteria and explain how you have reached your decision. The more fully this question is answered the more marks will be awarded.

SOLUTION

Assuming that the two samples were obtained under essentially the same type of conditions – same drivers, roads, weather conditions, speeds – let us look at the histograms.

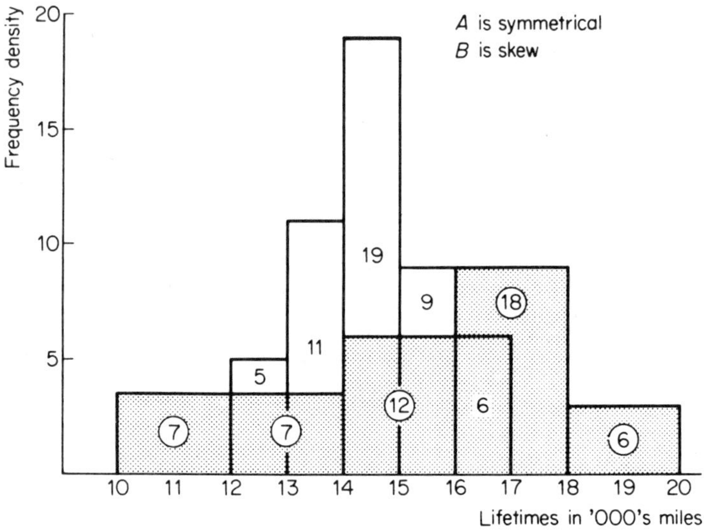

(*a*) The second type is more variable in performance.

(*b*) The second type has a higher mean – about 1000 miles on average.

(*c*) The two standard deviations are about 1000 miles and 2000 miles. (range/5).

(*d*) A 95 per cent confidence interval for the difference between the two means is:

$$1000 \pm 2\sqrt{\frac{100^2}{50}+\frac{2000^2}{50}}$$

$$10000 \pm 2000\sqrt{\frac{3}{50}}$$

$$1000 \pm 500 \text{ miles}$$

so the difference is real enough, between about 500 to 1500 miles. This is between

$$1/29 \times 100\% \text{ and } 1/9{\cdot}7 \times 100\%$$

a 3 to 10 per cent increase on average mileage.

If buying the tyres for a fleet, the saving on replacements would be about 5 per cent IF the uses were allowed to wear out the tyres, but obviously a tyre replacement policy would be used to minimise breakdown expenses.

Tyre *A* would be my choice IF the costs due to breakdown caused by tyre wear were high because they could be replaced with no breakdown costs at 12,000 miles.

Tyre *B* would be preferable if breakdown costs were minimal.

Normal

The normal distribution is a model which can be used in many business situations, and examiners seem keen to ensure that first-year students are familiar with the basic techniques for finding probabilities. One of the most straightforward types of problem is given below:

QUESTION

A nation-wide dairy company finds that the average number of pints of milk sold per roundsman per day is 1050, with a standard deviation of 380 pints. The distribution is normal.

(i) *What proportion of delivery men are selling less than 300 pints?*

(ii) *How many pints does a milkman need to sell to be within the top 10 per cent of delivery men?*

(iii) *If 100 milkmen are chosen at random, what is the probability that their average sales will exceed 1100 pints?*

SOLUTION

Drawing the normal curve will give a reference point for answering the question and provide rough estimates of the correct answer.

The automatic response then, to a question which mentions the normal distribution, is to draw a normal curve and ANNOTATE it:

(*a*) with areas corresponding to 1 & 2 standard deviations from the mean;
(*b*) with the name in full, of the variable;
(*c*) with the appropriate values of the variable.

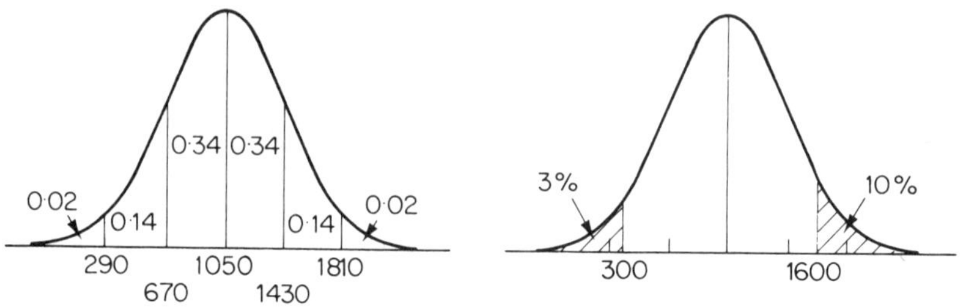

No. of pints of milk/roundsman/day

From the first rough diagram only about 3 per cent of delivery men sell less than 300 pints, and to be within the top 10 per cent a man must sell about 1650 pints.

More accurate answers can be found by using tables; 300 pints corresponds to a table value of about −2.

The exact value is

$$\frac{300 - 1050}{380} = -\frac{750}{380} = -1 \cdot 95$$

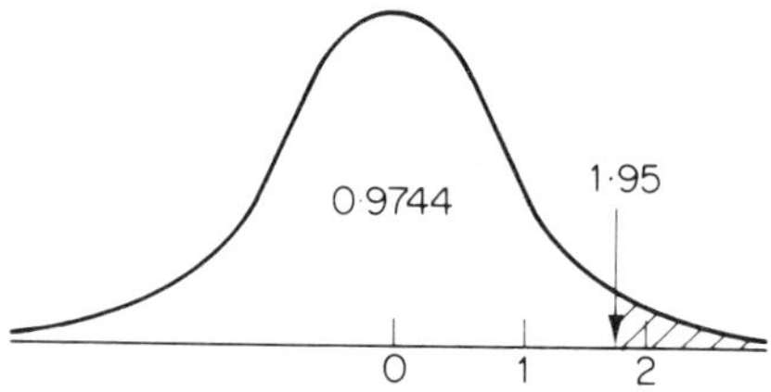

The value corresponding to 1·95 is 0·9744.
We require the tail area which is 1 − 0·9744 = 0·0256.
The proportion of delivery men selling less than 300 pints = 0·026.

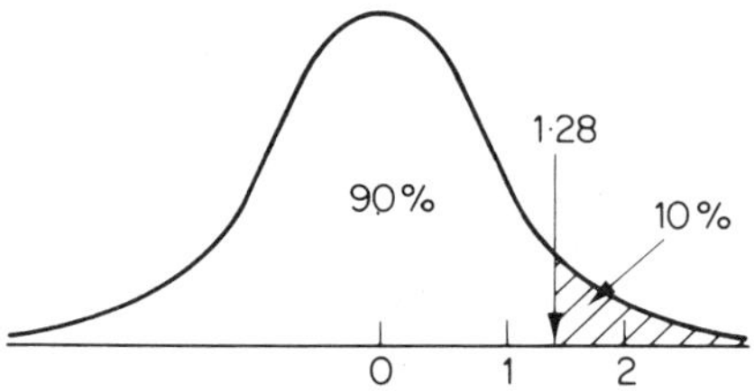

The table value to look up is 0·9 in the body of the table. The answer will be about 1·28 (standard deviations).

1·28 standard deviations is 394 pints, so a roundsman must sell 1444 (1050 + 394) pints to be in the top 10 per cent of delivery men.

The average sales of 100 milkmen is a variable. Take a different 100 milkmen and the average sales value will change.

Being a random variable, the average sales figure will have a mean and a standard deviation and a probability distribution.

These will be: a mean value of 1050; a standard deviation of $380/\sqrt{100} = 38$; a normal, or nearly normal probability distribution.

DRAW IT!

A value of 1100 will be over 1 standard deviation away from the mean so the required probability will be less than 0·16.

The table value for (1100 − 1050)/38 standard deviations is 0·9066.

The required probability is obtained by subtraction and is about 10 per cent 0·0934 to be exact.

If the data can be assumed to be reasonably normal, the properties of the cumulative distribution of the normal model can be used.

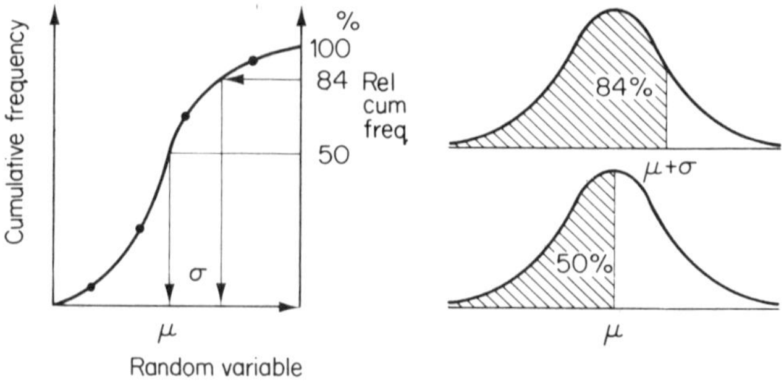

Given a set of frequencies reputedly from a normal population, the cumulative frequencies should be on a certain *S*-shaped curve. The point on the graph corresponding to the 50 per cent value will be the mean of population of values. The difference between the mean and the value corresponding to the 84 per cent value will be the standard deviation.

Drawing accurate *S*-curves through 4 or 5 points is not easy, but such is the usefulness of the technique that a special graph paper has been devised which extends the scale at the top and bottom of the figure while compressing it at the centre. *S*-shapes plotted on this paper are transformed to straight lines which are relatively easy to sketch.

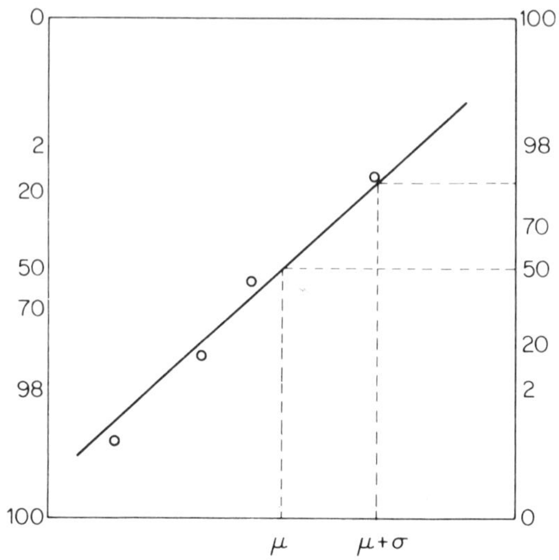

This method of estimating the mean and standard deviation is particularly requested in the next question.

QUESTION

A random sample of the weights to the nearest pound of 100 cases of bananas yields the following results:

Weight (lb)	*Frequency*
Less than 150	12
150–159	31
160–169	35
170–179	18
Greater than 179	4

Using probability graph paper estimate the mean and variance of the population. Hence or otherwise find:

(*a*) *Proportion of cases which exceed 150 lb.*
(*b*) *Proportion of cases between 145 and 170 lb.*
(*c*) *The tolerance limits which allow 90 per cent of the cases to be accepted.*

SOLUTION

The cumulative frequencies must be calculated first:

	c.f.	*Upper boundaries*
<150	12	150
150–159	43	160
160–169	78	170
170–179	96	180
>179	100	—

The plot on probability graph paper is approximately given below:

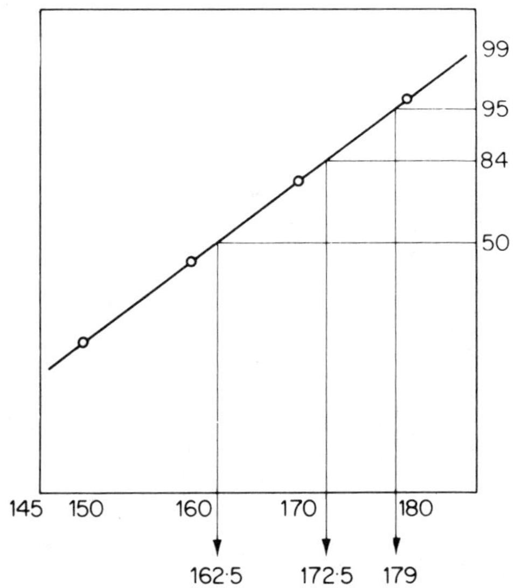

Reading the values from the graph (you should check these by drawing your own):

(*a*) $P(x > 150) = 0{\cdot}87$.
(*b*) $P(145 < x < 170) = 0{\cdot}82 - 0{\cdot}05 = 0{\cdot}77$.
(*c*) The middle 90 per cent of the values are between 145 and 179 lb. The mean is about 162·5 lb. The standard deviation is about $172{\cdot}5 - 162{\cdot}5$ lb $= 10$ lb.

Inference

Unfortunately, it is not possible to argue from the particular to the general and reach absolutely correct conclusions every time. Some statisticians would argue that particular results cannot *ever* be used to come to a definite conclusion; they can only modify the degree of belief that a person has about a proposition. Others maintain that by using certain statistical techniques the correct decision is made much more often than not. Statistical conclusions are therefore couched in probability terms and it is the conclusion which usually gives the most trouble in examination questions. Fortunately, if the answer follows a standard format the conclusion is a formality: it is a question of '. . . it doesn't matter what you do as long as you explain what it is you are doing'.

Consider the question:

QUESTION

(*a*) *In 1969 a random sample of 400 farm labourers showed that the average earnings was £17·14. The standard deviation from this sample mean was £4·48.*

(i) *Determine the 95 per cent confidence interval for average weekly earnings.*

(ii) *What do you understand by '95 per cent confidence interval'.*

(*b*) *A simple random sample of 144 primary classes in a local authority's schools contained 33 classes with 41 or more pupils. What proportion of primary classes in the local authority's area were in this category? Give your answer at the 95 and 99 per cent confidence interval.*

SOLUTION

The average earnings of 400 farm workers will be a random variable; we could call this anything so I will choose $\overline{W}$. $\overline{W}$ will have a probability distribution, a mean and a standard deviation and, if we assume that the interpretation of average is MEAN, then $\overline{W}$ will be normally distributed with a mean μ to be decided, and a standard deviation £4·48/$\sqrt{400}$ approximately.

The 95 per cent confidence limit for μ, t_L to t_u, is calculated from the probability statement about $\overline{W}$.

The probability that $\overline{W}$ lies between symmetrical limits t_L and t_u is 0·95.

$$p[t_L < \overline{W} < t_u] = 0{\cdot}95$$

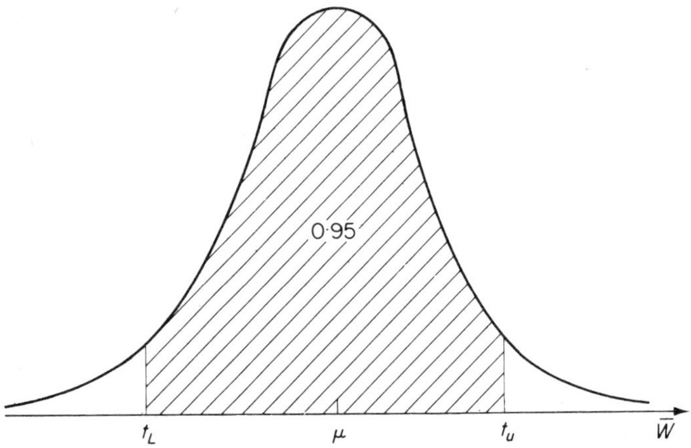

Distribution of $\overline{W}$

t_L and t_u will both be 1·96 standard deviations of $\overline{W}$ from μ, so the probability statement can be rewritten:

$$p\left[\mu - 1{\cdot}96\,\frac{4{\cdot}48}{\sqrt{400}} < \overline{W} < \mu + 1{\cdot}96\,\frac{4{\cdot}48}{\sqrt{400}}\right] = 0{\cdot}95$$

This is still a probability statement about $\overline{W}$. We can proceed one step further and construct a probability statement about a random interval defined as

$$\overline{X} \pm 1{\cdot}96\,\frac{4{\cdot}48}{\sqrt{400}}$$

The probability that this interval sits astride the unknown value of μ is 0·95. The implication is that IF data sets of size 400 were repeatedly selected at random from the population and the interval

$$\overline{X} \pm 1{\cdot}96\,\frac{4{\cdot}48}{\sqrt{400}}$$

calculated, 95 per cent of these intervals would straddle μ.

The 95 per cent confidence interval for μ is £17·14 ± 0·44.

QUESTION

(a) A customer is concerned about the goods he is buying from a certain manufacturer. The average weight of articles should be 8·50 grams, and on taking a sample of 10 from his stock he finds that the average weight is 8·26 grams with a standard deviation of 0·48 gram. Is the customer justified in complaining?

(b) A new system of processing and sending out orders claims to reduce the average waiting time of consumers for products. Samples taken before and after the system was introduced in one firm gave the following results.

Consumer waiting time	*No wait*	*Up to one week*	*Over one week*
Before new system	25	5	10
After new system	35	15	10

Using a χ^2 test show whether the new system has made a significant change at the 5 per cent level.

SOLUTION

Part (*a*) could be tackled by calculating a confidence interval but we will follow a hypothesis-testing approach. The first problem is to select a hypothesis to test.

We *could* test:

(*a*) *Ho:* the mean weight of articles IS 8·500 grams; or
(*b*) *Ho:* the mean weight of articles is 8·4 grams.

There are a number of hypotheses we could test but, since a figure of 8·5 grams is mentioned in the text of the question, let us assume that hypothesis (*a*) is true.

Now, carefully ignoring the fact that we know the value of the mean of the SAMPLE, we ask ourselves: If a sample of size 10 were taken from a normal (we had better assume normality I think) population of measurements and a statistic were to be calculated involving the mean and standard deviation of the sample values:

(*a*) What would be the distribution of the statistic?
(*b*) What would be considered a set of unlikely values of the statistic under *Ho*?

In this case the statistic in question will be

$$\frac{\overline{X} - \mu}{s/\sqrt{n}}$$

and it will have the '*t*' distribution.

A set of unlikely values would only have a small chance of being realised under *Ho*. This small chance value is a matter of taste and varies with the type of problem but with no other evidence to guide us, we can select 0·05 as a working value, bearing in mind it is completely arbitrary.

Where are these unlikely values to be found? There are several possibilities.

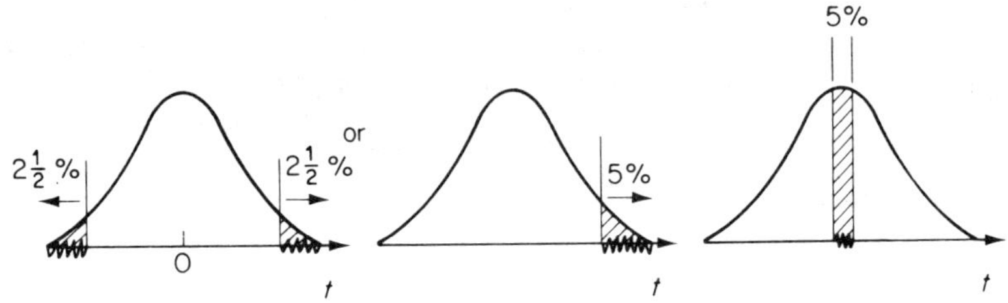

It seems that unlikely is a relative concept. We should ask 'unlikely compared with what ALTERNATIVE hypothesis?'

The clue to this alternative hypothesis is in the question. Presumably the customer would complain if the articles were bags of gold dust only if he received underweight goods, in which case the alternative hypothesis would be that $\mu < 8{\cdot}5$ grams. Under different circumstances, an overweight commodity could cause the customer to complain and if the commodity was claimed to have a standard weight, *any* deviation whether plus or minus could cause the customer to complain.

The problem of choosing the alternative hypothesis is not an easy one.

We will assume that the customer complains ONLY when supplied with underweight goods.

$$H_1 : \mu < 8{\cdot}5$$

If the statistic value falls in the critical region as defined below, *Ho* will be rejected in favour of H_1.

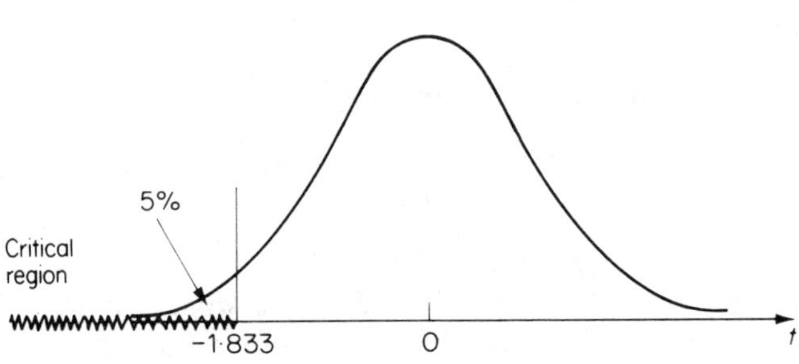

Now, and only now, after the arguing has been completed, do we calculate the value of the statistic

$$t = \frac{8\cdot26 - 8\cdot50}{\dfrac{0\cdot48}{\sqrt{10}}}$$

$$\text{assuming } s = \sqrt{\frac{\Sigma(x-\overline{x})^2}{9}}$$

Of course I could be wrong in the assumption about s. That $\sqrt{10}$ looks to be a nasty value to calculate. Maybe we should assume that

$$s = \sqrt{\frac{\Sigma(x-\overline{x})^2}{10}}$$

in which case

$$t = \frac{8\cdot26 - 8\cdot50}{\dfrac{0\cdot48}{\sqrt{9}}} = -\frac{(0\cdot24)\,.\,3}{0\cdot48} = -1\cdot5$$

This value does not fall in the critical region so *Ho* is not rejected in favour of H_1. The form of words for the conclusion could be: *Ho* was tested at the 5 per cent level of significance and was not rejected.

Apparently, it is not unreasonable to get a mean value for 10 items of 8·26 grams if the mean of all the values is 8·50 grams and the standard deviation is about 0·5 grams.

The customer is probably not justified in complaining.

(*b*) *Ho:* the proportions in the various categories are unchanged after the new system was introduced.

A skeleton table can now be constructed:

Consumer waiting time	*No wait*	*Up to one week*	*Over one week*	
Before new system	①	③	⑤	40
After new system	②	④	⑥	60
	60	20	20	100

If there was no difference in the various proportions, the best estimates of these proportions will be:

No wait 60/100
One week 20/100
One week+ 20/100

The actual values we would expect in the various positions ① to ⑥ would be:

① 60/100 . 40	③ 20/100 . 40	⑤ 20/100 . 40
② 60/100 . 60	④ 20/100 . 60	⑥ 20/100 . 60

The statistic is

$$\sum \frac{(\text{observed} - \text{expected})^2}{\text{expected}}$$

which has a χ^2 distribution with $(3-1)(2-1) = 2$ degrees of freedom. The critical value for χ_2^2 is 5·99.

A table of the components of the statistic is given below:

$\frac{(25-24)^2}{24}$	$\frac{(5-8)^2}{8}$	$\frac{(10-8)^2}{8}$
$\frac{(35-36)^2}{36}$	$\frac{(15-12)^2}{12}$	$\frac{(10-12)^2}{12}$

$$\chi_2^2 = \frac{1}{24} + \frac{9}{8} + \frac{4}{8} + \frac{1}{36} + \frac{9}{12} + \frac{4}{12} = \text{less than } 4$$

which is nowhere near 5·99.

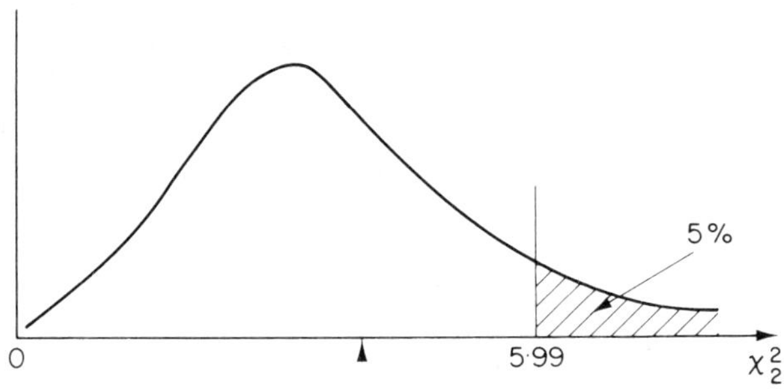

Distribution of χ_2^2

It is not possible to reject the hypothesis that the proportions of

no wait
up to 1 week
over 1 week

are independent of the type of system.

Put another way, the data does not provide sufficient evidence for us to say that the new system has significantly reduced waiting-time.

Note: The sample sizes were unequal. Why, do you think?

QUESTION

It is thought that two versions of a survey question should receive identical responses. The first version is applied to a random sample of 100 persons and the second to a random sample of 200 persons with the following results:

		VERSION		
		First	*Second*	TOTALS
Reply	YES	70	110	180
	NO	30	90	120
TOTALS		100	200	300

(*a*) *Briefly describe how you would test whether the difference in response rates was significant by using:*

(i) *A χ^2 (Chi-squared) test.*
(ii) *A test for the difference of two percentages.*

(*b*) *Use one of the tests and state clearly in words the meaning of your result.*

SOLUTION

We will consider (*b*).

If the responses to the question are supposed to be the same, what proportion of the target population would reply YES? We do not *know*, of course, but since 180 out of 300 replies were in the affirmative we could hypothesise that 180/300 of any sample from this homogenous target population would reply YES.

		VERSION			
		First	*Second*	TOTALS	
Reply	YES			180	$\frac{180}{300}$
	NO				
TOTALS				300	

We have before us two groups of 100 and 200. If these are just two samples from the *same* population, then we would *expect* a certain number of them to answer YES.

We would EXPECT 180/300 of the 100 in the first group to say YES, i.e. 60.

Table of EXPECTED values

First	*Second*
60	120
40	80

All we have to decide is whether the values in the original table are reasonably close to the expected values.

There is a statistic we can calculate which measures the percentage deviation[2]:

$$\chi^2 = \sum \frac{(\text{observed} - \text{expected})^2}{\text{expected}}$$

$$= \frac{(70-60)^2}{60} + \frac{(110-120)^2}{120} + \frac{(30-40)^2}{40} + \frac{(90-80)^2}{80} > 4$$

If the observed and expected values are close, χ^2 should be close to 0. How close is given by the tables as 3·84.

But $\chi^2 > 3{\cdot}84$. This value is unlikely to occur if the questions were the same. This value of χ^2 suggests that the two questions are being answered differently; they are not receiving identical responses.

Mathematics

Mathematics questions can be divided into two main types:

(1) The routine questions which ask for techniques to be demonstrated like the solving of an equation or the finding of a differential.

(2) Problems which express a mathematical model in words and require techniques to be applied to find a solution.

Questions of type 1 are easily recognised in past papers and must be rehearsed for success in the examination. Learning the techniques is not sufficient; they must be understood as well. Ask yourself:

Is the method I have learned the only way of doing it?
Does the method *always* work or are there exceptions?

If two methods can be used to solve the same problem you not only have a way of checking the answer but you have a better chance of relating that problem to other problems. The more links you build between different types of problem the easier mathematics becomes.

Type 2 questions need a different strategy for solution.

You should appreciate, first of all, that they are rarely solved by mathematicians in the manner of the model answer given at the end of the chapter. Their approach is more like that given below.

Initially, the variables in the problem are given letters. The relationships between variables are then written as equations and labelled so that they can be referred to easily.

For instance, the cost of producing television sets might be explained in terms of variables such as: overheads, tooling-up costs and material costs/TV.

The relationship would be:

Total cost = $H + T$ + cost of material for one TV × number made
i.e. $C = H + T + c \cdot n \ldots$ (1)

The next stage is to ask questions and try to find some answers. We could ask:

Are there values given for H, T, c and n?
How does this equation help solve the problem?
Does another equation contain H, T, c and n?
If so can we combine them?
Is the equation in its simplest form?

The equations are rearranged, simplified, extended, combined and shuffled like pieces in a jigsaw puzzle. Mistakes are made and possible solution lines come to dead ends. Sometimes the work proceeds from what is given until no further progress can be made and so a start is made assuming a solution and working backwards until the missing piece is identified and fitted in. The whole process is like piecing together a jigsaw. Once you have stumbled on the way to do it, reassembling it again is much easier. The model answer is the method of reassembly.

QUESTION

A monopolist faces a demand curve (a), *and a total cost curve* (b).

(*a*) $P = 60 - 2q$ ①
(*b*) $TC = 10 + 20q$ ② } *label the equations!*

where P *is the price per unit of the product (in pounds),* q *is the amount produced and sold in units, and* TC *is the total cost in pounds.*

(i) *At what output are profits maximised?*

(ii) *If a tax of* t *pounds per unit is levied on the monopolist how does the profit-maximising output vary with* t*?*

(iii) *What tax rate would produce the maximum amount of tax revenue for the government?*

(iv) *What would the firm's output be if that tax rate were imposed?*

SOLUTION

This is a pure manipulation problem with a smattering of elementary economics thrown in to disguise the pure algebra.

(i) The profits are to be maximised, so the profits against quantity-produced curve must have a bump in it. None of the given equations is other than pure linear so a profit equation must be constructed.

$$\text{Profit} = \text{revenue} - \text{cost}$$
$$= (\text{number produced})(\text{cost/unit}) - \text{cost}$$
$$\text{Profit} = qP - TC$$

Profit must now be made a function of q alone:

$$\text{Profit} = q(60 - 2q) - 10 + 20q \quad \textit{from (1) and (2)}$$
$$\text{Profit} = 80q - 2q^2 - 10 \quad \textit{simplifying}$$

We could now plot this for $q = 0, 1, \ldots, 20$ and find the largest value for profit and the corresponding value for q but, remembering elementary calculus, the maximum will occur when

$$\frac{d(\text{Profit})}{dq} = 0$$

$$\frac{d(\text{Profit})}{dq} = 80 - 4q$$

when $80 - 4q = 0$, $q = 20$ *maximum occurs when* $\dfrac{dP}{dq} = 0$

Profits are maximised when the quantity produced is 20.

(ii) Profit now has to be redefined.

$$\text{Profit} = \text{revenue} - \text{cost} - \text{tax} \quad \textit{a word-relationship}$$
$$\text{Tax} = qt$$
$$\text{Profit} = 80q - 2q^2 - 10 - qt$$

Differentiating again and setting $d(\text{Profit})/dq = 0$, we have $80 - 4q - t = 0$, and the quantity required for maximum profit is $(80 - t)/4$.

(iii) The tax revenue $= qt$.

This will be maximised when q is a maximum but the manufacturer has decided that q will be $(80 - t)/4$ so

$$\text{Tax revenue} = \left(\frac{80 - t}{4}\right) . t$$

$$\frac{d(\text{Tax})}{dt} = \frac{80 - 2t}{4}$$

At a maximum

$$\frac{d(\text{Tax})}{dt} = 0 \qquad \therefore t = 40$$

Tax revenue is maximised when the tax rate is £40.

Index